T0202195

The New Statistics with R

The New Statistics with R

An Introduction for Biologists

Second Edition

ANDY HECTOR

Department of Plant Sciences and Linacre College,
University of Oxford, UK

OXFORD
UNIVERSITY PRESS

OXFORD
UNIVERSITY PRESS

Great Clarendon Street, Oxford, OX2 6DP,
United Kingdom

Oxford University Press is a department of the University of Oxford.
It furthers the University's objective of excellence in research, scholarship,
and education by publishing worldwide. Oxford is a registered trade mark of
Oxford University Press in the UK and in certain other countries

First Edition published in 2015

Impression: 1

Published in the United States of America by Oxford University Press
198 Madison Avenue, New York, NY 10016, United States of America

British Library Cataloguing in Publication Data

Data available

Library of Congress Control Number: 2021931174

ISBN 978-0-19-879817-0 (hbk.)
ISBN 978-0-19-879818-7 (pbk.)

DOI: 10.1093/oso/9780198798170.001.0001

Printed in Great Britain by
Bell & Bain Ltd., Glasgow

I dedicated the first edition of this book to the memory of Christine Müller.
This new edition is dedicated to Lindsay and Rowan.

Acknowledgements

The original version of this book was begun at the end of 2011 while I was on sabbatical as a visiting researcher in the computational ecology group at Microsoft Research in Cambridge—my thanks to Drew Purves and colleagues for their support. This second edition was partly written during my sabbatical in 2019/20, sadly largely under covid-19 restrictions. However, before lockdown I made some important progress during stays at Obertschappina—thanks Roland and Petra—and on a visit to the Cedar Creek Ecosystem Science Reserve—for which I thank Forest Isbell, Dave Tilman, and the amazing group of ecologists at the University of Minnesota.

Several people were instrumental in helping cultivate my initial interest in statistical analysis. I was first introduced to experiments during my final-year project with Phil Grime and colleagues at the UCPE at Sheffield University. Shortly afterwards, one of the most rewarding parts of my PhD at Imperial College was learning statistics (and the GLIM software) from Mick Crawley. Bernhard Schmid shared this interest and enthusiasm and taught me a lot while I was a postdoc on the BIODEPTH project and, later, when we worked together at the Institute for Environmental Sciences at the University of Zurich (sorry for forsaking Genstat for R, Bernhard!). Here in Oxford I have continued to discuss and learn about statistics partly through the generosity of Geoff Nicholls.

I have also benefited from sometimes brief but important discussions with several other statisticians during training courses, after visiting talks, and the like, including Douglas Bates, Andrew Gelman (over a game of Quincunx), Martin Maechler, Peter McCullagh, John Nelder, José Pinheiro, Bill Venables, and Hadley Wickham. My apologies to them for any misunderstandings that make it into this book.

Many group members helped me delve further into statistics with R, including some of the material covered in this book. I would like to thank all current and past group members, but particularly Robi Bagchi, Stefanie von Felten, Yann Hautier, Charlie Marsh, Chris Philipson, Matteo Tanadini, Sean Tuck, Maja Weilenmann, and Mikey O'Brien. I have also learned a lot from collaborating on papers on statistics with several colleagues, including Tom Bell, Jarrett Byrnes, John Connelly, Laura Dee, Forest Isbell, Marc Kéry, Michel Loreau, and Alain Zuur.

The content of this book is based on teaching materials developed over the last two decades at Imperial College, the University of Zurich, and here at Oxford, where I teach statistics at the Bachelor, Masters, and PhD levels. Thanks to everyone involved—particularly the many demonstrators (TAs).

Many people helped find errors in the first edition of this book—I have tried to correct them and acknowledge the spotters at the R café website (no doubt there will be more to add for this second edition). In particular, my thanks to Ben Bolker for his constructive criticism of the first edition of this book.

At OUP, thanks go to Ian, Lucy, Bethany, and Charlie for making this book and this second edition possible. Also, thanks to Douglas Meekison who has skilfully copyedited the manuscript and Sumintra Gaur has been project manager for this book.

Finally, thank you—and sorry—to anyone who has slipped my mind as I rush again to meet the book deadline!

Andy Hector, Oxford, October 2020.

Contents

Chapter 1: Introduction   **1**

  1.1   Introduction to the second edition   1
  1.2   The aim of this book   2
  1.3   Changes in the second edition   3
  1.4   The R programming language for statistics and graphics   4
  1.5   Scope   4
  1.6   What is not covered   5
  1.7   The approach   5
  1.8   The new statistics?   6
  1.9   Getting started   6
  1.10 References   7

Chapter 2: Motivation   **9**

  2.1   A matter of life and death   9
  2.2   Summary: Statistics   12
  2.3   Summary: R   13
  2.4   References   13

Chapter 3: Description   **15**

  3.1   Introduction   15
  3.2   Darwin's maize pollination data   16
  3.3   Summary: Statistics   28
  3.4   Summary: R   28
  3.5   References   28

Chapter 4: Reproducible Research   **29**

  4.1   The reproducibility crisis   29
  4.2   R scripts   30

4.3   Analysis notebooks   32
4.4   R Markdown   32
4.5   Summary: Statistics   37
4.6   Summary: R   37
4.7   References   37

Chapter 5: Estimation   **39**

5.1   Introduction   39
5.2   Quick tests   40
5.3   Differences between groups   41
5.4   Standard deviations and standard errors   43
5.5   The normal distribution and the central limit theorem   45
5.6   Confidence intervals   48
5.7   Summary: Statistics   50
5.8   Summary: R   50
       Appendix 5a: R code for Fig. 5.1   50

Chapter 6: Linear Models   **51**

6.1   Introduction   51
6.2   A linear-model analysis for comparing groups   52
6.3   Standard error of the difference   57
6.4   Confidence intervals   58
6.5   Answering Darwin's question   60
6.6   Relevelling to get the other treatment mean and standard error   62
6.7   Assumption checking   63
6.8   Summary: Statistics   66
6.9   Summary: R   67
6.10   Reference   67
       Appendix 6a: R graphics   67
       Appendix 6b: Robust linear models   68
       Appendix 6c: Exercise   68

Chapter 7: Regression   **71**

7.1   Introduction   71
7.2   Linear regression   72
7.3   The Janka timber hardness data   73
7.4   Correlation   75
7.5   Linear regression in R   75

7.6	Assumptions	78
7.7	Summary: Statistics	82
7.8	Summary: R	83
7.9	Reference	83
	Appendix 7a: R graphics	83
	Appendix 7b: Least squares linear regression	84

Chapter 8: Prediction — 85

8.1	Introduction	85
8.2	Predicting timber hardness from wood density	85
8.3	Confidence intervals and prediction intervals	90
8.4	Summary: Statistics	94
8.5	Summary: R	95

Chapter 9: Testing — 97

9.1	Significance testing: Time for t	97
9.2	Student's t-test: Darwin's maize	98
9.3	Summary: Statistics	106
9.4	Summary: R	106
9.5	References	106

Chapter 10: Intervals — 107

10.1	Comparisons using estimates and intervals	107
10.2	Estimation-based analysis	108
10.3	Descriptive statistics	109
10.4	Inferential statistics	113
10.5	Relating different types of interval and error bar	119
10.6	Summary: Statistics	124
10.7	Summary: R	125
10.8	References	125

Chapter 11: Analysis of Variance — 127

11.1	ANOVA tables	127
11.2	ANOVA tables: Darwin's maize	128
11.3	Hypothesis testing: F-values	132
11.4	Two-way ANOVA	135
11.5	Summary	137
11.6	Reference	138

Chapter 12: Factorial Designs   **139**

12.1 Introduction   139
12.2 Factorial designs   139
12.3 Comparing three or more groups   142
12.4 Two-way ANOVA (no interaction)   145
12.5 Additive treatment effects   148
12.6 Interactions: Factorial ANOVA   152
12.7 Summary: Statistics   158
12.8 Summary: R   159
12.9 References   159
      Appendix 12a: Code for Fig. 12.3   160

Chapter 13: Analysis of Covariance   **161**

13.1 ANCOVA   161
13.2 The agricultural pollution data   162
13.3 ANCOVA with water stress and low-level ozone   165
13.4 Interactions in ANCOVA   171
13.5 General linear models   172
13.6 Summary   175
13.7 References   176

Chapter 14: Linear Model Complexities   **177**

14.1 Introduction   177
14.2 Analysis of variance for balanced designs   178
14.3 Analysis of variance with unbalanced designs   180
14.4 ANOVA tables versus coefficients: When F and t can disagree   184
14.5 Marginality of main effects and interactions   186
14.6 Summary   192
14.7 References   192

Chapter 15: Generalized Linear Models   **195**

15.1 GLMs   195
15.2 The trouble with transformations   196
15.3 The Box–Cox power transform   200
15.4 Generalized Linear Models in R   203
15.5 Summary: Statistics   208
15.6 Summary: R   208
15.7 References   208

Chapter 16: GLMs for Count Data **209**

16.1 Introduction 209
16.2 GLMs for count data 210
16.3 Quasi-maximum likelihood 213
16.4 Summary 215

Chapter 17: Binomial GLMs **217**

17.1 Binomial counts and proportion data 217
17.2 The beetle data 218
17.3 GLM for binomial counts 220
17.4 Alternative link functions 225
17.5 Summary: Statistics 228
17.6 Summary: R 228
17.7 Reference 228

Chapter 18: GLMs for Binary Data **229**

18.1 Binary data 229
18.2 The wells data set for the binary GLM example 230
18.3 Centering 236
18.4 Summary 238
18.5 References 238

Chapter 19: Conclusions **239**

19.1 Introduction 239
19.2 A binomial GLM analysis of the *Challenger* binary data 239
19.3 Recommendations 246
19.4 Where next? 249
19.5 Further reading 249
19.6 The R café 249
19.7 References 250

Chapter 20: A Very Short Introduction to R **251**

20.1 Installing R 251
20.2 Installing RStudio 253
20.3 R packages 254
20.4 The R language 254

Index 259

Introduction

1.1 Introduction to the second edition

Back in 2015, I opened the introduction to the first edition of this book as follows:

> Unlikely as it may seem, statistics is currently a sexy subject. Nate Silver's success in out-predicting the political pundits in the last US election drew high-profile press coverage across the globe (and his book many readers). Statistics may not remain sexy but it will always be useful. It is a key component in the scientific toolbox and one of the main ways we have of describing the natural world and of finding out how it works. In most areas of science, statistics is essential.

So much has changed over the last five years. Initially, I thought this introduction to the second edition would discuss the subsequent failure of statistics to predict the Brexit referendum and Trump election results. However, I ended up working on this second edition under lockdown due to the COVID-19 pandemic. I'm not sure if statistics is still 'sexy' but it is certainly still prominent in our lives. Modelling, much of it statistical, provides predictions of the spread of COVID-19, and sampling is key to estimating fundamental parameters like the reproductive number, denoted

The New Statistics with R: An Introduction for Biologists. Second Edition. Andy Hector, Oxford University Press. © Andy Hector 2021. DOI: 10.1093/oso/9780198798170.003.0001

(coincidentally) *R*—the number of people each person with COVID-19 in turn infects.

1.2 The aim of this book

This book is intended to introduce one of the most useful types of statistical analysis to researchers, particularly in the life and environmental sciences: linear models and their generalized-linear-model (GLM) extensions. My aim is to get across the essence of the statistical ideas necessary to intelligently apply and interpret these models in a contemporary ('new') way. I hope it will be of use to students at both undergraduate and postgraduate levels and to researchers interested in learning more about statistics (or in switching to the software packages used here, R and RStudio). The approach is therefore not primarily mathematical, and makes limited use of equations—they are easily found in numerous statistics textbooks and on the internet if you want them. I have also kept citations to a minimum and give them at the end of the most relevant chapter (there is no overall bibliography). The approach is to learn by doing, through the analysis of real data sets. That means using a statistical software package, in this case the R programming language for statistics and graphics (for the reasons given below). It also requires data. In fact, most scientists only start to take an interest in statistics once they have their own data. In most science degrees that comes late in the day, making the teaching of introductory statistics more challenging. Students studying for research degrees (Masters and PhDs) are generally much more motivated to learn statistics since they know it will be essential for the analysis of their data. The next best thing to working with our own data is to work with some carefully selected examples from the literature. I have used some data from my own research but I have mainly tried to find small, relevant data sets that have been analysed in an interesting way (preferably by a qualified statistician). Most of them are from the life and environmental sciences. I am very grateful to all of the people who have helped collect these data and developed the analyses (they are named in the appropriate chapters as the data and example are introduced). For convenience, I have tried to use data sets that are available within the R software.

1.3 Changes in the second edition

The first edition of this book was written following standard procedure to supply a Word document of the text of each chapter plus files of any figures. This proved an inefficient and error-prone method with all the copy–pasting between R scripts and the word processing file. This second edition has been entirely rewritten using the R Markdown package to produce a PDF file of each chapter along with the TeX file that generates it (as I understand it, subcontractors will then use LaTeX to apply the book format). Writing the second edition like this should be a smarter, more efficient, and hopefully less error-prone way to work. In the process, the book has changed in many ways. Based on my experience in teaching the Quantitative Methods for Biology course at Oxford, the content has been divided up into a greater number of bite-size topics that will hopefully prove more digestible for students and more useful to teachers. In part because the book was written using the R Markdown package, I now drive R using the RStudio software (it also provides a standard interface on all platforms and lots of other great support materials, like the R cheat sheets). Every chapter has been rewritten but there are also entirely new chapters, one giving an opening motivational example, one on reproducible research (using the R Markdown package), and another on some of the complexities of linear-model analysis that I skipped over in the first edition. There are now separate chapters on GLMs for the analysis of different types of non-normal data. The first edition also contained chapters on mixed-effects and generalized linear mixed-effects models (GLMMs). These have been dropped from the second edition—partly due to the space limits but also because some reviewers and readers felt that one chapter was just not enough even for a short introduction to mixed-effects models. Furthermore, the example GLMM no longer ran using later versions of the software.

1.4 The R programming language for statistics and graphics

R is now the principal software for statistics, graphics, and programming in many areas of science, both within academia and outside (many large companies use R). There are several reasons for this, including:

- R is a product of the statistical community: it is written by the experts.
- R is free: it costs nothing to download and use, facilitating collaboration.
- R is multiplatform: versions exist for Windows, Mac, and Linux.
- R is open-source software that can be easily extended by the R community.
- R is statistical software, a graphics package, and a programming language all in one (as we'll see, you can now even produce books, blogs, and websites from R).

1.5 Scope

Statistics can sometimes seem like a huge, bewildering, and intimidating collection of tests. To avoid this I have chosen to focus on the linear-model framework as probably the single most useful part of statistics (at least for researchers in the environmental and life sciences). The book starts by introducing several different variations of the basic linear-model analysis (analysis of variance, linear regression, analysis of covariance, etc.). I then introduce an extension: generalized linear models for data with non-normal distributions. The advantage of following the linear-model approach is that a wide range of different types of data and experimental designs can be analysed with very similar approaches. In particular, all of the analyses covered in this book can be performed in R using only two main functions, one for linear models (the lm() function) and one for GLMs (the glm() function), together with a set of generic functions that extract different aspects of the results (confidence intervals etc.).

1.6 What is not covered

This book is primarily about statistics (linear models), not the R software. For that, OUP offers introductory volumes by Beckerman et al. (2017) and Petchey et al. (2021). Statistics is a huge subject, so the limited size of the book precluded the inclusion of many topics, and the coverage is limited to linear models and GLMs. There was no space for non-linear regression approaches, generalized additive models (GAMs). Because of the focus on an estimation-based approach, I have not included non-parametric statistics. Experimental design is covered briefly and integrated into the relevant chapters. The use of information criteria and multimodel inference are briefly introduced. The basics of Bayesian statistics is also a book-length project in its own right (e.g. Korner-Nievergelt et al. 2017).

1.7 The approach

There are several different general approaches within statistics (frequentist, Bayesian, information theory, etc.) and there are many subspecies within these schools of thought. Most of the methods included in this book are usually described as belonging to 'classical frequentist statistics'. However, this approach, and the probability values that are so widely used within it, has come under increasing criticism. In particular, statisticians often accuse scientists of focusing far too much on P-values and not enough on effect sizes. This is strange, as the effect sizes—the estimates and intervals—are directly related to what we measure during our research. I don't know any scientists who study P-values! For that reason, I have tried to take an estimation-based approach that focuses on estimates and confidence intervals wherever possible. Styles of analysis vary (and fashions change over time). Because of this, I have tried to be frank about some of my personal preferences used in this book. In addition to making wide use of estimates and intervals, I have also tried to emphasize the use of graphs for exploring data and presenting results. I have tried to encourage the use of *a priori* contrasts (comparisons that were planned in advance) and I advocate avoiding the inappropriate overuse of multiple testing in favour of a more focused, planned approach. Finally, at the end of each chapter I try to summarize both the statistical approach and what it has enabled us to learn about the science of each example. It is easy to get lost in statistics, but for non-statisticians the analysis should not become an end in its own right, only a method to help advance our science.

1.8 The new statistics?

What is the 'new statistics' of the title? The term is not clearly defined but it appears to be used to cover a combination of new techniques—particularly meta-analysis—with a back-to-basics focus on estimation-based analysis using confidence intervals (Cumming 2012). Meta-analysis is beyond the scope of this edition—I recommend the book by Koricheva et al. (2013). In this book, the 'new statistics' refers to a focus on estimation-based analysis, together with the use of modern maximum-likelihood-based analysis (including information criteria) plus methods for reproducible research. I have also tried to take account of the recent criticisms of the overuse of *P*-values and statistical significance (although this is an area of ongoing debate).

1.9 Getting started

To allow a learning-by-doing approach, the R code necessary to perform the basic analysis is embedded in the text along with the key output from R (files of the R code will be available as support material from the R café at http://www.plantecol.org/). Some readers may be completely new to R, but many will have some familiarity with it. Rather than start with an introduction to R, we will dive straight into the example analyses. However, a brief introduction to R is provided at the end of the book, and newcomers to the software will need to start there.

1.10 References

Beckerman, A., Childs, D.Z., & Petchey, O.P. (2017) *Getting Started with R.* Oxford University Press.

Cumming, G. (2012) *Understanding the New Statistics.* Taylor and Francis.

Koricheva, J., Gurevitch, J., & Mengersen, K. (2013) *Handbook of Meta-analysis in Ecology and Evolution.* Princeton University Press.

Korner-Nievergelt, F., Roth, T., von Felten, S., Guélat, J., Almasi, B., & Korner-Nievergelt, P. (2017) *Bayesian Data Analysis in Ecology Using Linear Models with R, BUGS, and Stan.* Academic Press.

Petchey, O.L., Beckerman, A.P, Cooper, N., & Childs, D.Z. (2021) *Insights from Data with R: An Introduction for the Life and Environmental Sciences.* Oxford University Press.

Silver, N. (2012) *The Signal and the Noise.* Penguin.

Motivation

2.1 A matter of life and death

The Space Shuttle *Challenger* (Fig. 2.1) was one the most advanced spacecraft ever built, but the first version lacked ejection seats for its crew. This was particularly relevant in January 1986, when the expected temperature at launch was below freezing, much colder (around 30 °F) than on any previous mission, raising concerns over safety. The shuttle (*Challenger* was one of only five put into service) was designed to be largely reusable. The orbiter—the main shuttle craft—was propelled out of the atmosphere by its own engines (supplied by a large external liquid-fuel tank that could be jettisoned when empty) and with the help of two booster rockets that disengaged when spent and fell back, to be recovered from the sea for reuse. The booster rockets were constructed in cylindrical sections and the joints sealed with huge circular washers called O-rings. It was these O-rings that were of particular concern, as their ability to prevent fuel leaks ('blow by' in NASA jargon) depended on their flexibility and plasticity, which decreased as temperatures fell.

The New Statistics with R: An Introduction for Biologists. Second Edition. Andy Hector,
Oxford University Press. © Andy Hector 2021. DOI: 10.1093/oso/9780198798170.003.0002

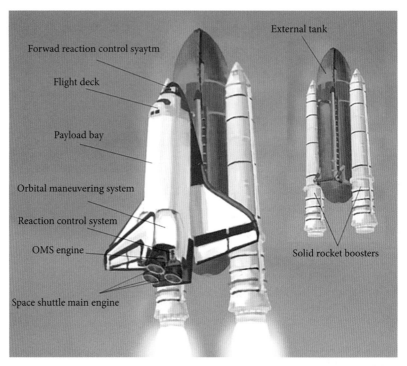

Figure 2.1 A schematic diagram of the space shuttle, showing the orbiter with external liquid-fuel tank and reusable booster rockets. Copyrighted free use, https://commons.wikimedia.org/w/index.php?curid=554970

Because the boosters were recovered and refurbished for reuse, it was possible to infer from scorch marks whether fuel leak damage had occurred during each launch. The resulting data (called orings) is available as part of an R package called faraway (think of packages as add-on apps that you can download to extend the 'base' version of R—this package accompanies the book by Faraway (2014)). The 'chunk' of R code below activates the faraway package (it must already be installed—see Chapter 20) and displays the orings data (the head() function shows only the first several rows to save space—note that the lines of R output are prefixed with two hashes to distinguish them):

```
library(faraway)
head(orings)
```

```
##      temp damage
## 1     53       5
## 2     57       1
## 3     58       1
## 4     63       1
## 5     66       0
## 6     67       0
```

A graph of the number of leaks (from launches prior to *Challenger*'s 1986 mission) as a function of launch temperature looks like that shown in Fig. 2.2 (the library() function loads the ggplot2 package so that its quick-plot function can use the orings data to draw a scatterplot with temperature on the *x*-axis and damage on the *y*, saving the graph as 'Fig2_2'):

```
library(ggplot2)
Fig2_2 <- qplot(data = orings, x = temp, y = damage)
Fig2_2
```

Do you think the number of fuel leaks is related to temperature?

A teleconference was held between NASA and the booster rocket manufacturer on the eve of the launch and, after prolonged discussion, the decision was made to proceed. Tragically, shortly after lift-off a fuel leak

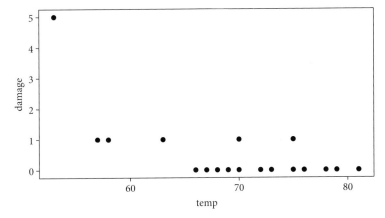

Figure 2.2 The relationship between the number of fuel leaks ('damage') and launch temperature for shuttle launches prior to *Challenger*'s 1986 mission.

on one of the booster rockets ignited the fuel in the external tank, causing an explosion that destroyed *Challenger* and killed all seven crew members.

As the well-known physicist and author Richard Feynman famously demonstrated at the subsequent inquiry using a cup of ice water (Tufte 2005), it was the unusually low temperature that caused the O-rings to become brittle, allowing fuel to leak and causing the explosion. One revelation was that the discussion between NASA and the engineering company did not involve any statistical analysis of the relationship between fuel leaks and temperature on previous launches—they did not even look at simple graphs of the data like those shown above. You might think that to analyse the available data, to at least draw graphs, is not exactly rocket science—and even if it was, these were rocket scientists!

The moral is that we should never underestimate the importance of statistical analysis—including informal graphical analysis. In this case, it could have saved the lives of seven people. Fortunately, most analyses do not have such important consequences, but statistics is widely used in medicine, health and safety, and other areas where lives are also at stake. Even when the outcomes are less critical, graphical and statistical analysis is a key part of research. Unfortunately, we are in the middle of a reproducibility crisis—many scientific results that are published cannot be later repeated, due to problems with the research process, including the analysis. While we are scientists, not statisticians, it is essential to use statistics as part of our research and to do so as well as possible and to improve on the current state of affairs.

2.2 Summary: Statistics

In this motivating chapter we haven't conducted a formal statistical analysis. However, a graphical analysis is usually a good way to begin, as we did here. We won't pursue the *Challenger* example any further now (as we'll see later in the book, analysis of this type of data is too advanced to make a good starting point). Instead we will begin with a simpler type of linear-model analysis and some biological data from none other than Charles

Darwin—who was a keen early experimenter. The data are also found in an R package, in this case called SemiPar (from now on the R packages needed are given at the beginning of each chapter).

2.3 Summary: R

- R has a minimal 'base' distribution that can be supplemented by many add-on packages.
- Functions are a fundamental part of the R language; they have names followed by a pair of parentheses.
- Once installed, R packages are activated for use with the library() function.
- R is an object-oriented programming language—the assignment arrow '->' or '<-' is used to create R objects.

2.4 References

Faraway, J.J. (2014) *Linear Models with R*. CRC Press.
Tufte, E. (2005) *Visual Explanations*. Graphics Press.

Description

3.1 Introduction

We are going to begin with a linear-model analysis where the aim is to compare the effects of different treatments. Each treatment is applied to a group of experimental units and if the treatments are effective they will produce differences between the groups. Experiments are designed to test hypotheses about the effects of the treatments and to quantify the strength of these effects. In this chapter we will focus on describing the treatment effects (a step we often skimp on in our haste to get to the punchline), saving testing for later.

3.1.1 R PACKAGES

This exercise requires the following R packages (which are often loaded again at first use to indicate where they are needed):

```
library(ggplot2)
library(Sleuth3)
library(SMPracticals)
```

The New Statistics with R: An Introduction for Biologists. Second Edition. Andy Hector, Oxford University Press. © Andy Hector 2021. DOI: 10.1093/oso/9780198798170.003.0003

3.2 Darwin's maize pollination data

The loss of genetic diversity is an important issue in the conservation of species that have declined in their population sizes due to over exploitation, habitat fragmentation, and other causes. Scientific interest in inbreeding depression dates back to at least Charles Darwin, who wrote an entire book on the subject, including many results from his own work—Darwin was a keen early experimenter (Costa 2017). In *The Effects of Cross and Self-Fertilisation in the Vegetable Kingdom*, Darwin (1876) describes how he produced seeds of maize (*Zea mays*) that were fertilized with pollen from the same individual or from a different plant. Pairs of seeds taken from self-fertilized and cross-pollinated plants were then grown in pots and the height of the young seedlings was measured as a surrogate for their evolutionary fitness. Darwin wanted to know whether inbreeding reduced the fitness of the selfed plants—this is the biological hypothesis the experiment was designed to test. Darwin asked his cousin Francis Galton— an early statistician (and, infamous as originator of eugenics)—for advice on the analysis. At that time, Galton could only lament that

> The determination of the variability is a problem of more delicacy than that of determining the means, and I doubt, after making many trials whether it is possible to derive useful conclusions from these few observations. We ought to have measurements of at least fifty plants in each case.

Luckily, we now have a variety of ways of determining the variability, even from smaller sample sizes, that allow us to answer Darwin's question.

A full version of Darwin's data can be found in the SMPracticals package—the dataframe (R terminology for a data set) is called darwin. The R language works with objects (R is an object-oriented programming language)—among other things, these can be single values (scalars), columns of numbers (vectors), dataframes, or graphs (see Chapter 20). R objects, including this dataframe, can be seen simply by typing their name (try it), but to save space the head() function shows just the first few rows:

```
library(SMPracticals)
head(darwin)
```

```
##    pot pair  type height
## 1    I    1 Cross 23.500
## 2    I    1  Self 17.375
## 3    I    2 Cross 12.000
## 4    I    2  Self 20.375
## 5    I    3 Cross 21.000
## 6    I    3  Self 20.000
```

3.2.1 KNOW YOUR DATA

The first thing to do with any data set is to explore it in order to understand its structure in relation to the experimental design. There are R functions for exploring data that we will use every time we have a new data set. The structure function, str(), explains that the dataframe has 30 observations of 4 variables (30 rows × 4 columns):

```
str(darwin)
```

```
## 'data.frame':    30 obs. of  4 variables:
## $ pot   : Factor w/ 4 levels "I","II","III",..: 1 1 1 1 1 1 2 2 2 2 ...
## $ pair  : Factor w/ 15 levels "1","2","3","4",..: 1 1 2 2 3 3 4 4 5 5 ...
## $ type  : Factor w/ 2 levels "Cross","Self": 1 2 1 2 1 2 1 2 1 2 ...
## $ height: num  23.5 17.4 12 20.4 21 ...
```

The first three variables are *factors*—categorical variables that divide the data into different groups. The groups are called *levels*, and in this case the factors have 4, 15, and 2 levels that are indicated with roman numerals, numbers, and words, respectively. The final variable is numeric—it is a continuous measure of height. The summary() function gives a summary of each column's contents—how many experimental units (maize plants) are in each of the factor levels and some basic summary statistics for plant heights:

```
summary(darwin)
```

```
##    pot            pair         type          height
##  I  : 6    1        : 2    Cross:15    Min.   :12.00
##  II : 6    2        : 2    Self :15    1st Qu.:17.53
##  III:10    3        : 2                Median :18.88
##  IV : 8    4        : 2                Mean   :18.88
##            5        : 2                3rd Qu.:21.38
##            6        : 2                Max.   :23.50
##            (Other):18
```

Take some time to understand the output of the structure() and summary() functions in relation to the design of the experiment and the layout of the dataframe.

3.2.2 SUMMARIZING AND DESCRIBING DATA

Darwin measured plant height (in inches, here in decimal form) as a surrogate for fitness—this is our *response variable*. The experimental treatment applied two types of hand pollination, either with pollen from the same plant ('selfed') or from another plant ('crossed'); this is our *explanatory variable*. We want to explain responses in plant height as a function of differences in the experimental pollination treatment. There are 15 plants of each type, planted in pairs, which is indicated by a factor with 15 levels (one for each pair). Darwin planted the pairs of maize plants in large pots (we'll ignore this for now and come back to it later). The SMPracticals darwin dataframe is in long (or 'tidy') format: every experimental measurement has its own row and every variable has its own column. This is almost always the best format for data and it is the form the data must be in for analysis using the R linear-model function, lm(), that we are going to use for the analysis.

However, before we start the analysis using the lm() function we want to perform some simple calculations. Students arriving at university from school are used to doing calculations 'by hand' and so we'll start in this vein by using R as an oversized pocket calculator. The functions we will use for the basic calculations need the data in wide format instead (with the selfed and pollinated plants in each pair side by side). R has many tools for rearranging data, but we can postpone that complexity for now as the Sleuth package (version 3 or the earlier version 2) already has the data in wide format in example 428, ex0428 (note that the information on plant pots is not given):

```
library(Sleuth3)
head(ex0428)
```

```
##    Cross  Self
## 1 23.50 17.38
## 2 12.00 20.38
## 3 21.00 20.00
## 4 22.00 20.00
## 5 19.13 18.38
## 6 21.50 18.63
```

We can plot the data using the quick-plot function, qplot(), from the Grammar of Graphics package, ggplot2, that we have already met briefly. A ggplot2 graph must be drawn using information contained in a single data set (or dataframe, as R calls it), specified using the data argument. We can then specify which variable in the darwin dataframe is the explanatory (x) variable and which is the response (y) variable (we can improve the graph by adding things like full axis labels, but we'll postpone that until later to keep the R code as short and simple as possible for now). Recall that R is an object-oriented programming language—we work by creating objects (rather than having to type them out again and again), which we can then use and modify as needed. In this case, we create a ggplot2 graph object (using the assignment arrow) that we've called fig3_1, which we can then view by typing its name:

```
fig3_1 <- qplot(data = darwin, x = type, y = height)
fig3_1
```

The graph, shown in Fig. 3.1, suggests that the average height may be greater for the crossed plants, which would be consistent with a negative

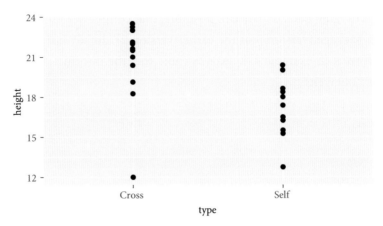

Figure 3.1 Individual plant heights as a function of pollination type.

effect of inbreeding. But how confident can we be in this apparent signal of inbreeding depression given the level of noise created by the variation within groups? The variability seems reasonably similar in the two groups except that the crossed group has a low value lying apart from the others with a height of 12—a potential *outlier*. We'll look at this in more depth later.

Above, we have produced graphs that plot all of the raw data. This is often the best strategy with data sets that have groups with small numbers of data points in them, where calculating summary statistics (like standard deviations) could produce unrepresentative results skewed by extreme values. However, when the number of data points per group is not small it can be useful to use graphs that plot summary statistics. The well-known statistician John Tukey devised box-and-whisker plots (boxplots) that show the median values (a measure of central tendency that splits the data in half), the first and third quartiles, which contain the middle half of the data, and whiskers containing 95% of the data plus individual outlying values. Medians are a useful measure of central tendency for data with skewed distributions, where mean values can be unrepresentative (which is why we often hear about median rather than mean salaries, for example). We can modify our quick-plot R code using an argument to change the *geom*, short for 'geometric object' (the symbols, lines, etc.):

```
fig3_2 <- qplot(data = darwin, x = type, y = height, geom = "boxplot")
fig3_2
```

The result is shown in Fig. 3.2.

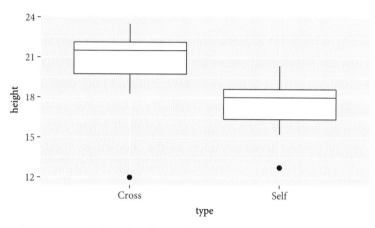

Figure 3.2 Boxplot of plant height as a function of pollination type.

Another alternative is a violin plot (geom = violin) (Fig. 3.3), which in our case does a good job of visualizing the higher and lower 'centres of gravity' of the distributions of the crossed and selfed data, respectively:

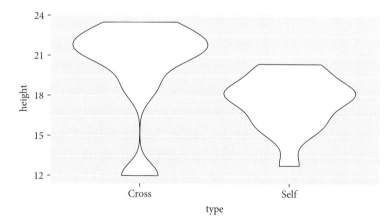

Figure 3.3 Violin plot of plant height as a function of pollination type.

```
fig3_3 <- qplot(data = darwin, x = type, y = height, geom = "violin")
fig3_3
```

Perhaps the most popular options—barcharts or bar plots—have drawbacks and are usually not as good as the better alternatives shown above.

A recent popular book about statistics by Nate Silver (2012), called *The Signal and the Noise*, nicely sums up the primary goals of our analysis: we want to find out if there is any systematic pattern in our data (signal) that stands out above the background variability (noise). We usually quantify the signal in terms of the average value for each group (although other measures of central tendency such as the median or mode are more appropriate in some cases, as briefly discussed above). Before we calculate the mean for each group we can start more simply, ignoring pollination type, and do the same thing for all of the height measurements. R has a mean() function, which we can use to get the average height of all 30 plants. We need to tell R where to find the column of numbers to use for the calculation. There are a few different ways to do that. One approach is to use the with() function, which has a data argument to indicate the name of the dataframe we want to use:

```
with(data = darwin, mean(height))
```

```
## [1] 18.88333
```

So, the average height of all 30 plants is around 18.9 inches.

Estimating the average response is straightforward, but now we come to the problem Galton met: how to quantify the variability. We do so using a series of related quantities. We begin with the *variance*. The variance is calculated through a process called least squares, but we'll put an explanation of that to one side for the moment and return to it later. The variance is also known as the mean squares and is referred to as S^2 when we are talking about a sample of data, as here, or symbolized as sigma-squared,

σ^2, when we are referring to the population from which the sample is drawn (although in books on applied statistical analysis these abbreviations and symbols are sometimes used more loosely, a crime I shall no doubt also commit here!). R has a var() function for calculating the variance. Again, we need to specify the dataframe. Another way to do this is to indicate a column within the dataframe by giving the name of the dataframe and the variable separated by a dollar sign, like$this (it is sometimes more convenient to use the with() function and sometimes the dollar notation— I will use both, since you will encounter both frequently in books and in other people's code):

```
var(darwin$height)
```

```
## [1] 10.11846
```

However, there's a catch: our estimate of the variability is not on the same scale as the original data. Instead, we have a measure of variation on the squared scale. This leaves us with a measure of the signal in inches and a measure of the noise in square inches (something we would normally associate with an area rather than a height!). The solution is straightforward— we simply reverse the earlier squaring. Taking the square root of the variance gives us a measure of variability on the same scale as the original data: the standard deviation, abbreviated to SD and symbolized as S and as a lowercase Greek sigma, σ (when referring to the whole population)):

$$SD = \sqrt{s^2}$$

Strictly, the Latin letter refers to the sample standard deviation and the lowercase Greek sigma to the population standard deviation, but they are sometimes used more loosely. R has the sd() function for calculating the standard deviation:

```
sd(darwin$height)
```

```
## [1] 3.180953
```

The standard deviation is the average difference between an individual measurement (individual plant height) and the mean value. We can check its relationship to the variance using the sqrt() function (note the nesting of one R function inside another—make sure you have equal numbers of opening and closing brackets, a common mistake in code):

```
sqrt(var(darwin$height))
```

```
## [1] 3.180953
```

3.2.3 COMPARING GROUPS

We have used the mean(), var(), and sd() functions to calculate the overall mean, variance, and standard deviation for the heights of all 30 plants from the darwin dataframe (ignoring the grouping into self- and cross-pollinated plants to begin with). We can do the same for the two experimental treatments using the Cross and Self columns in the wide-format data set (the ex0428 dataframe from the Sleuth package shown above):

```
mean(ex0428$Cross)
## [1] 20.19333
mean(ex0428$Self)
## [1] 17.57667
```

You can replace the mean() function with sd() to get the measures of variability (and so on to calculate other summary statistics for the two groups).

While this approach works fine here as an easy way to get started, it does not scale well for larger, more complex data sets. As usual in R, there are various alternative solutions but they usually require the data in long (tidy) format, so we need to switch back to using the darwin dataframe. The base version of R provides the tapply() (table-apply) function to calculate summary statistics for each group. The arguments for the tapply() function are the name of the variable we want to perform the calculation on, the name of the factor that defines the groups we want to calculate summary statistics for (the indexing factor), and the type of summary statistic we want to produce (the function). Here are the means...

```
with(data = darwin, tapply(height, type, mean))
```

```
##      Cross      Self
## 20.19167 17.57500
```

......and the standard deviations:

```
with(data = darwin, tapply(height, type, sd))
```

```
##      Cross      Self
## 3.616945 2.051676
```

We now have the numbers we need for a basic description of the signal and noise in our data—the means and the standard deviations for the two groups. When possible, it is usually better to present numbers primarily as graphs rather than as tables (you can always supplement the graph with a table when exact values are needed). As usual in R, there are many ways to draw this graph. To try and keep things simple and transparent, we will make a small table of the summary statistics, which qplot() will then turn into a graph for us (we'll look at alternatives with more complex R code later). We can take our lines of R code for calculating the means and SDs and use the assignment arrow to create variables (objects) to save their values, which we can then use to draw the graph (I won't do so here, to save space, but it's often a good idea to check what you've created, which you can do by typing the object's name to view it on screen):

```
means <- tapply(darwin$height, darwin$type, mean)
sds <- tapply(darwin$height, darwin$type, sd)
```

It is safer not to use terms used by R (e.g. function names like mean and sd) when naming objects, to avoid the obvious potential confusion.

Although the summary statistics will automatically inherit the names of the factor levels (here Cross and Self), we can also create columns of text— for example, we can create a third variable to give full names to the two treatment types:

```
pollination <- c("Crossed", "Selfed")
```

The information we need to draw a graph with the ggplot2 package must be included in a single dataframe (the dataframe name is given as the first argument). We can combine our three variables into a dataframe of summary statistics for Darwin's data (which we've called dar_sum_stats):

```
dar_sum_stats <- data.frame(pollination, means, sds)
dar_sum_stats
```

```
##         pollination    means       sds
## Cross      Crossed 20.19167 3.616945
## Self        Selfed 17.57500 2.051676
```

We want a graph that shows our measures of central tendency (the means in this case) as points and our measure of variability (the standard deviations) as lines ('error bars' or 'intervals'—as we'll see later, there are many different types). The ggplot2 package has a wide selection of geometric objects ('geoms') we can use to draw this type of graph, but many of them share arguments (ymin and ymax) which we can use to tell R the values for the ends of the 'error bar' lines. In this case, we specify the ymax and ymin values to be the means plus and minus one standard deviation:

```
fig3_4 <-
  qplot(data = dar_sum_stats, x = pollination, y = means) +
  geom_pointrange(aes(ymin = means - sds, max = means + sds)) +
  theme_bw()
fig3_4
```

So far, we have taken the default grey background for the ggplot2 graphs but a range of other options are available, including the black and white

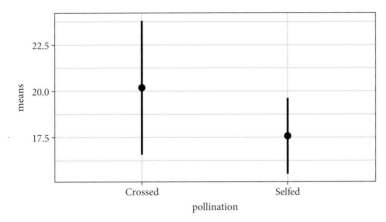

Figure 3.4 Mean plant height as a function of pollination type (±1 standard deviation).

	pollination	means	sds
Cross	Crossed	20.19167	3.616945
Self	Selfed	17.57500	2.051676

Figure 3.5 Table produced by kable() function.

theme used here in Fig. 3.4 (now that you know how to do this, I shall mostly set this to be the default for the book to save ink).

Graphs are usually a better way to communicate information than tables, but if we want to accompany our figure with a supplementary table then the kable() function is a simple way to convert the monospaced-font version of dar_sum_stat given above into something of publication quality (Fig. 3.5):

```
kable(dar_sum_stats)
```

Descriptive statistics are often a step analysts skimp on in their haste to get to testing their question of interest. We've seen various descriptive graph types (plots showing all of the data, boxplots, and violin plots) and ended here with the classic measures of central tendency and variability for samples of this type—means and standard deviations. The tables and graphs show that the mean height of the cross-pollinated plants is greater than that of the selfed ones and the variability is greater too—this pattern, where the variability increases with the mean, is a common one that we will explore further later.

3.3 Summary: Statistics

Description is an essential early phase of all statistical analysis. The starting point for an analysis is the experimental design. The aim is to use one or more explanatory variables to understand differences in a response variable. Categorical explanatory variables are known as factors. We can describe the signal with a measure of central tendency (mean, median, or mode as appropriate) and the noise with a measure of dispersion (usually the standard deviation).

3.4 Summary: R

- Data sets are called dataframes in R jargon.
- Always use the summary() and str() functions to get a thorough initial understanding of data before analysis.
- R has a wide range of packages and functions for generating descriptive graphs of data.

3.5 References

Costa, J.T. (2017) *Darwin's Backyard*. Norton.

Darwin, C.R. (1876) *The Effects of Cross and Self-Fertilisation in the Vegetable Kingdom*. Echo Library (2007).

Reproducible Research

4.1 The reproducibility crisis

One of the key features of scientific research is that results should be reproducible, and the same goes for the statistical analyses that the results are based on. Unfortunately, over the last decade or so we have discovered that a worrying amount of our work is not reproducible (or replicable—these terms are sometimes used for different aspects of the problem but are also often used interchangeably). The causes of this reproducibility crisis are complex and act at many points of the research process, but improving how we go about our statistical analysis will certainly help. Point-and-click statistical software may be quick to start working with, but could you ever repeat the exact sequence of clicks you have made? No—to do that we need a computer program, which is exactly what we have if we save a copy of our R code as an R script. Scripting is a huge advance over point-and-click, but we can go further and use the R Markdown package to produce 'analysis notebooks' to better record and explain what we have done and why. The aim of this chapter is to make a short example of a script and then to convert it to an R Markdown document. Due to space

The New Statistics with R: An Introduction for Biologists. Second Edition. Andy Hector, Oxford University Press. © Andy Hector 2021. DOI: 10.1093/oso/9780198798170.003.0004

limitations, the example is deliberately minimal and just demonstrates how to combine (knit) together some short input code, R output (output text, figures, and tables), and some narrative text. To avoid the need for extra packages it uses only those that come with the base distribution of R, including the plot() function from the base graphics package rather than using ggplot2. Luckily, the authors of the R Markdown package have written books about it (using R Markdown, or rather its offshoot R Bookdown), that are freely available on line in web page form (Xie et al. 2019; 2021).

4.1.1 R PACKAGES

```
library(ggplot2)
library(SMPracticals)
```

4.2 R scripts

An R script is just a text file that saves all of the R commands used in an analysis from first to last. One drawback of R scripts is that there is limited scope to add explanatory text: we can only include one-line comments and these must be prefixed with the hash symbol, #. For example, the following R code applies the summary() function to the co2 dataframe—a time series of atmospheric carbon dioxide concentrations:

```
summary(co2)
```

```
##     Min. 1st Qu.  Median    Mean 3rd Qu.    Max.
##    313.2   323.5   335.2   337.1   350.3   366.8
```

The (base R) plot() function draws a line graph of the information it contains on changes in atmospheric concentrations of carbon dioxide over time (Fig. 4.1):

```
Fig4_1 <- plot(co2)
```

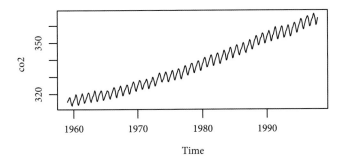

Figure 4.1 Example Figure: Atmospheric carbon dioxide concentrations 1959–1997 (ppm).

A minimal script of the combined lines of R code with some comments would look something like this:

```
# Title: Atmospheric carbon dioxide concentrations
# Author: Andy Hector Summary statistics for
# concentrations of atmospheric CO2
summary(co2)
# Line graph of changes in atmospheric carbon dioxide
# (1959 - 1997)
plot(co2)
```

To write an R script, we need to stop writing commands directly into the console (where they are lost once we close R)—instead, we can open a new window for the script using the RStudio menu options File -> New File -> R script. The result is just a text file, but saving it from within RStudio will assign it a .R extension. Double-clicking it will open it within R or RStudio rather than a default text file reader (you can specify which application you want to open files with the .R extension within the system preferences—we are using RStudio, so set it to that rather than R). Saving an R script allows us, or anyone else, to reproduce an analysis (just as the methods in a scientific paper should allow others to recreate our scientific studies) so long as we make both the script and the data available. However, readers still need both the script and the data, and have to run the script (this could be time-consuming with a complex analysis and a big data set). A complementary approach is to use the R script to produce a *reproducible research document* that guides the reader through the analysis process.

4.3 Analysis notebooks

The reproducibility crisis suggests that our current methods of analysis leave a lot to be desired. One problem is the poor documenting of how the analysis was done. In some ways, it is odd how casual we are about documenting our analyses. For example, in areas of science where the details of the research could be examined in court (as part of a patent dispute for example), keeping detailed lab notebooks is compulsory, sometimes with very strict protocols for how they must be filled out. Why not apply similar standards to how data analysis is performed? Writing scripts is a big contribution to reproducible research, but we can go further and produce 'analysis notebooks' that document the analysis process in a similar way to that in which lab notebooks record what was done at the bench. The R Markdown package currently provides the easiest way to generate reproducible research documents from R (R Markdown uses the earlier knitr package, which followed the even earlier sweave). Outside of R, popular options include Jupyter notebooks. The idea is to produce documents that knit (weave) together the input code and the software-generated output (text, tables, and figures) with our own text to produce an understandable narrative of the analysis process from start to finish. As a brief example, let's take the example R script given above and convert it into an analysis notebook using R Markdown.

4.4 R Markdown

R Markdown documents are produced by the package of the same name (in conjunction with other R packages and software). The R Markdown package is included with RStudio, so we should not need to install the package—it will be used when needed (we don't even need to use the library() function to activate it). A new R Markdown document can be opened using the menu options File -> New file -> R Markdown, which will open a template for a new R Markdown document where you can give a title and author (HTML is the recommended default choice) (Fig. 4.2).

The template that opens is not blank—it attempts to help explain how R Markdown works. Unfortunately, it is more complex than it needs to be. At the top is a header (in YAML) that simply reproduces the information you entered when creating the file a moment ago. I suggest you delete everything below the header (which starts and ends with lines of three dashes) so that your document looks like Fig. 4.3.

One of the big differences between an R Markdown document and a script is that in an R Markdown document the R code is segregated into 'chunks'. You can insert a new empty chunk (ready to hold R code) using

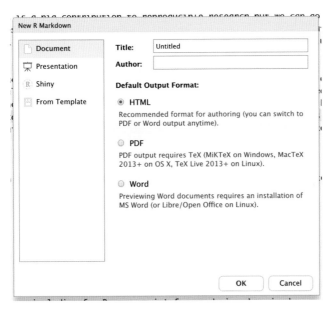

Figure 4.2 Creating a new R Markdown document in RStudio.

```
1 ▾ ---
2   title: "Atmospheric carbon dioxide concentrations"
3   author: "Andy Hector"
4   date: "14/04/2020"
5   output: html_document
6   ---
7
```

Figure 4.3 The R Markdown document header.

the menu options Code -> Insert Chunk (or click the green +CInsert icon in the upper right corner). R chunks stand out from the rest of the document in grey shading. Now simply paste in the first line of R code from the R script. Because the code is segregated into a chunk, we can add as much text outside the chunk as we like without having to use hashes to distinguish text from code. If we add a second chunk for the plot() command and convert the comments to text, the R Markdown document should look like Fig. 4.4.

The R code in a chunk can be run by clicking the green arrow on the upper right-hand side. If the R code produces output, it is shown directly underneath the chunk. Our example now looks like Fig. 4.5 (if it does not, then click on the gear symbol at the top of the screen and choose Chunk Output Inline).

We now have a document on screen that integrates R input and output together with our narrative text. To produce a file of the document, click on the Knit button at the top of the screen and choose 'knit to html_document'.

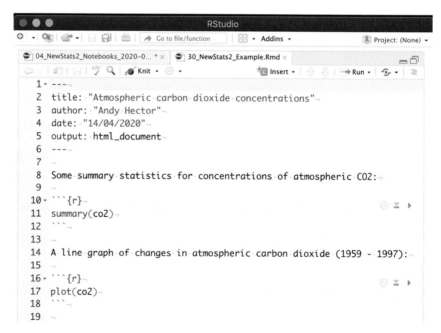

Figure 4.4 The R Markdown document separates (grey) chunks of R code from the main narrative text.

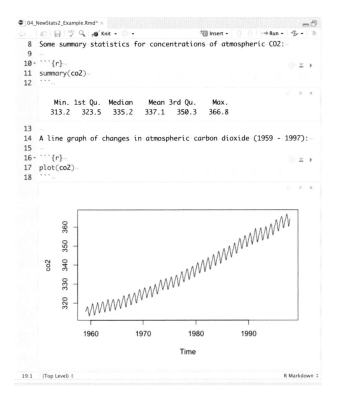

Figure 4.5 On-screen preview of our R Markdown analysis notebook in RStudio.

It is possible to generate many other file types. If you have Microsoft Word installed then that should appear as an output option in the Knit menu. To produce PDFs you will need a version of LaTeX. One option is to install MacTeX on macOS—https://tug.org/mactex/—or MiKTeX on Windows—http://miktex.org—but running the following line of code should install the tinytex package, which should allow you to produce PDFs (Fig. 4.6):

```
tinytex::install_tinytex()
```

There are many advantages of working this way. It is much more efficient and less error-prone than copy-pasting R output into other software (as I did to write the first edition of this book). And the key point in the context of this chapter is that our work is reproducible. R Markdown documents

Atmospheric carbon dioxide concentrations

Andy Hector

14/04/2020

Some summary statistics for concentrations of atmospheric CO2:

```
summary(co2)
```

```
##    Min. 1st Qu.  Median    Mean 3rd Qu.    Max.
##   313.2   323.5   335.2   337.1   350.3   366.8
```

A line graph of changes in atmospheric carbon dioxide (1959 - 1997):

```
plot(co2)
```

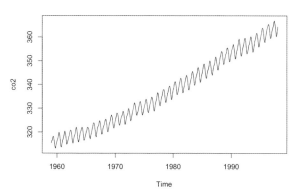

Figure 4.6 The knitted html document.

are *dynamic documents*—any changes to the data or R code automatically propagate through the analysis workflow to produce updated output. These 'analysis notebooks' are mostly for our own use, documenting every step from start to finish, showing all R input and output. However, it is possible to specify options in the chunk headings that allow you more control over what is shown. For example, we may want to show a graph but not the R code that produced it, and this is easy to do (see the R Markdown cheat sheet and reference guide available via the RStudio help menu for a summary, and Xie et al. (2019) for longer explanations). This allows you to produce a broader range of documents—*reproducible research reports* if you like—for sharing with a wider group of readers, including those who don't use R.

One final point. R is updated every 6 months. For many classical analyses this won't change anything, but for newer approaches that are still in development—mixed-effects models for example—results can and do change from version to version. There are more sophisticated ways, but the sessionInfo() function is a quick way of recording what version of R was used for the analysis (and other details of the operating system of the computer used, etc.). This allows users to recreate an analysis even when the software has changed, because old versions of R are archived (on CRAN). Running this line of code (remember to add it as a final chunk to your R Markdown file) will produce output that details the software versions used:

```
sessionInfo()
```

4.5 Summary: Statistics

Scientific results should be reproducible. A key part of reproducible research is the use of dynamic documents. Currently, the easiest way to produce dynamic documents in R is to use the R Markdown package that comes as part of RStudio. Further information on R Markdown can be found in the cheat sheets available via the RStudio Help menu and in Xie et al. (2019; 2021).

4.6 Summary: R

- The SessionInfo() function produces a record of the software versions used for an analysis.

4.7 References

Xie, Y., Allaire, J.J., & Grolemund, G. (2019) *R Markdown: The Definitive Guide*, CRC Press.

Xie, Y., Dervieux, C., & Riederer, E. (2021) *R Markdown Cookbook*, CRC Press.

Estimation

5.1 Introduction

Statistics is all about signal and noise. In the analysis of Darwin's maize data we have focused initially on description: we quantified the signal in Darwin's maize data in the form of measures of central tendency (mean and median values), and the noise using the standard deviations. But what Darwin wanted to know was whether pollination affects fitness—specifically, whether self-pollination is detrimental. In other words, our question concerns the *differences* in height—something we have not yet explicitly looked at. In this chapter we will focus on estimating the differences in height and testing whether we think there is an effect of the pollination treatments. Our goal is to estimate the mean heights of the two treatments in Darwin's experiment and the difference between them, and to calculate various measures that quantify our *confidence* (and the flip side, *uncertainty*) in the estimates, which we can use to judge whether they appear to be different or not.

The New Statistics with R: An Introduction for Biologists. Second Edition. Andy Hector, Oxford University Press. © Andy Hector 2021. DOI: 10.1093/oso/9780198798170.003.0005

5.1.1 R PACKAGES

```
library(arm)
library(ggplot2)
library(Sleuth3)
library(SMPracticals)
```

5.2 Quick tests

It is very easy for non-statisticians to lose sight of the big picture and get lost in the details of the maths and programming when doing analysis. The highly creative statistician John Tukey invented a 'quick' test that practitioners could use without needing a computer (this was back in the late 1950s). Let's take a look at Darwin's maize data again to get the gist of Tukey's quick test (the R code for the graph is given at the end of the chapter) (Fig. 5.1).

The figure displays the heights of the maize plants for the two groups side by side from highest to lowest. First, take a look at the tops of the two stacks of data points. The tallest maize plant is one of the cross-pollinated

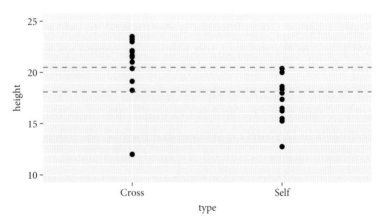

Figure 5.1 Plant height as a function of pollination type; the red line shows the tallest selfed plant, and the blue line the shortest crossed one (ignoring low outlying values).

group, with a height of nearly 24 inches, while the largest selfed individual only just makes it to around 20 inches (shown by the horizontal dashed red line). Even a quick look at the graph makes it clear that several of the cross-pollinated plants are taller than the largest self-pollinated plants. However, it turns out that the shortest plant is also cross-pollinated but that this individual is much shorter than any other cross-pollinated plant—it is what we can call an *outlier*. It's tempting to ignore (or even remove) outliers, but that would be cheating! However, for the sake of exploring Tukey's quick test let's pretend for a moment that this outlying value does not exist. If that were the case then the shortest cross-pollinated plant would be around 18 inches (dashed blue line) and there would be at least half a dozen shorter self-pollinated plants. Tukey's quick test test works by counting up how many cross-pollinated plants exceed the height of the tallest self-pollinated plant and how many selfed plants are shorter than the lowliest crossed plant (where 7 roughly corresponds to the conventional minimum level for starting to have confidence that there is an effect). In other words, to what degree is one stack of data points shifted up relative to the other? (Ideally, we'd like the stacks of points to be of similar distance from top to bottom, that is, to have similar variability within groups.) I have been deliberately vague about the numbers here as we can't apply Tukey's method in full due to the outlying values, but the upward shift of the sample of heights of the crossed plants relative to the selfed ones is consistent with Darwin's hypothesis of a negative effect of inbreeding on fitness. If we ignored the shortest cross-pollinated plant, Tukey's quick test would strongly support Darwin's hypothesis. What does a more formal test say if we include all of the data?

5.3 Differences between groups

Darwin's data consist of matched pairs of seedlings. When data are matched it is better to work with the differences in height, rather than the means for the treatments (ignoring the matching). Let's first calculate the differences in height for each of the 15 pairs of cross- and self-pollinated plants.

A useful feature of R is that it does vector-based calculations. For example, we can subtract one column of numbers from another to calculate the difference in height for each pair. Again, note the use of the dollar sign to indicate dataframe$variable, and how we can use this here to create the new column of the differences in height *inside* the ex0428 dataframe where we want it (otherwise it would be homeless in R's workspace):

```
ex0428$Difference <- ex0428$Cross - ex0428$Self
head(ex0428)
```

```
##    Cross  Self Difference
## 1 23.50 17.38       6.12
## 2 12.00 20.38      -8.38
## 3 21.00 20.00       1.00
## 4 22.00 20.00       2.00
## 5 19.13 18.38       0.75
## 6 21.50 18.63       2.87
```

We can now describe the differences in height by calculating the mean difference:

```
mean(ex0428$Difference)
```

```
## [1] 2.616667
```

... and the standard deviation:

```
sd(ex0428$Difference)
```

```
## [1] 4.719373
```

The standard deviation is intended for describing the level of variation in a sample, not for testing how confident we are that a difference between samples exists. For that we need to go a bit further and calculate two related measures: the *standard error* and the *confidence interval*.

5.4 Standard deviations and standard errors

One potentially confusing issue is the difference between the standard deviation and the standard error—after all, the names are similar, as are the formulas (see below). As we've seen, the standard deviation is a descriptive statistic—it measures the variability (*dispersion*) in a sample of data. The standard deviation applies to the individual data points— it's the average distance between a data point and the mean. However, we now want to move beyond description to inference: given the size of the difference, and the level of variability, can we infer that the groups are different? To do this we need to shift our focus from the individual data points to the difference between the means. Instead of a measure of variability of the sample of data points we need one for the mean, and that is what the standard error is. We can think of the standard error as a standard deviation for the mean. After all, our mean is just an estimate based on this one sample of data—repeating the same experiment with another 30 plants would produce a different sample and a different estimate of the mean. The standard error describes the variability we would expect among samples if we repeated the experiment over and over again, calculating a new mean each time (the sampling distribution of the means). The standard error is related to the standard deviation as

$$ SE = \frac{s}{\sqrt{n}} $$

and to the variance as

$$ SE = \sqrt{\frac{s^2}{n}} $$

where s is the sample standard deviation (sigma, σ, is used when dealing with an entire population), s^2 is the sample variance (again, σ^2 is used for the population), and n is the sample size (the number of values used in the calculation). The measure of variability on the top of the equation is divided by the sample size on the bottom. As the sample size goes up, the standard error gets smaller, reflecting the decreasing uncertainty (and our increasing confidence) in the estimate when it is based on larger samples. This makes good intuitive sense—just think about how much trust you put in poll predictions based on the number of people interviewed. However, the relationship is one of diminishing returns—the standard deviation is divided by the square root of the sample size, so that while larger samples produce estimates with lower uncertainty, to halve the uncertainty we need to quadruple the sample size (and so on).

If we continue to work with the single sample of the 15 paired differences in plant height, we can calculate a standard error for the mean difference in height as follows:

```
sd(ex0428$Difference)/sqrt(15)
```

```
## [1] 1.218537
```

A 'point estimate' (like our mean) is of limited use without a measure of our precision for the estimated value (the standard error). Point estimates should always be presented with a measure of precision, for example

. . . the difference in height was 2.62 ± 1.22 inches (mean ± SE).

Now that we have measures of the effect size and its precision, how can we use the standard error to infer our level of confidence in whether there is a difference between means or not? To do so, we can set up a null hypothesis. In this case our null hypothesis is one of no difference between means—in other words, a difference of zero. Our estimated mean difference in height is 2.62 inches—this is our putative signal. We need to assess our confidence or, looking at it from the other direction, our uncertainty in this signal given the level of background noise. The standard error provides our estimate of the uncertainty. The larger the signal relative to the noise, the more confident we can be that it is real (we can never be 100% sure, of course—we can never completely rule out the possibility of a false positive). The question is: how likely are we to have estimated a difference of 2.62 if the null hypothesis was correct and there was actually zero difference? To quantify how likely it is we can use a probability distribution—in this case the normal (or Gaussian) distribution.

5.5 The normal distribution and the central limit theorem

As you will hopefully recall from school, the normal distribution is the well-known bell-shaped curve. The normal distribution is defined by two *parameters*—the mean and the standard deviation. The value of the mean determines the position where the bell curve is centred (the value of its highest point). The distribution is symmetric about this point, with the width of the bell—the length of its tails—determined by the value of the standard deviation. Large values of the standard deviation produce short, wide bells with long tails and small values generate tall, narrow curves with short tails. The bell curve is well known because it occurs so frequently. This is because this is how things often vary when they are influenced by many small effects—a common occurrence. For example, human heights are usually approximately normally distributed because height is determined by the effects of many genes and environmental influences during development. As with all probability distributions, the area under the bell curve sums to one and known proportions of the curve lie within certain distances from the centre of the bell curve. The centre is defined by the mean and the distance from the centre is measured in standard deviations. The idealized *standard normal distribution* (Fig. 5.2) is defined to have a mean of zero and a standard deviation of one. Just over two-thirds of the area under the curve (67.8%) lies within ±1 SD, 95% within ±2 SDs, and 99.8% within ±3 SDs.

But, how does this help us work out how likely we are to observe a difference of 2.62 inches in plant height if the 'true' difference is zero? The idea—thanks to the central limit theorem—is to use the normal distribution as a model for the variability in our sample of data. We are going to assume that the properties that characterize the normal distribution will hold approximately true for our sample. It is important to understand that

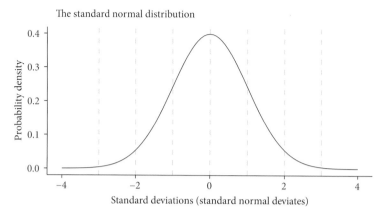

Figure 5.2 The standard normal distribution. Dashed lines mark ±1, 2, and 3 standard deviations around the mean of zero.

any inferences we draw using this approach depend on the assumptions being satisfactorily met—something we need to check (we will postpone that until we repeat this analysis using the R linear-model function lm()). It is also important to remember that using the properties of the normal distribution as a model for the variability in a sample of data works less well as sample sizes become small (as we'll see later, we can then switch to an alternative to the normal distribution that takes account of sample size). Putting these two issues to one side, we can now construct a bell curve that takes the values for its parameters from our sample (Fig. 5.3). The middle of the bell curve is centred on the estimated mean difference in height—a value of 2.62. Just over two-thirds of the area under the curve (68%) lies within ±1 standard error of the mean (±1.22 inches), 95% of it within ±2 standard errors, and almost all of it (99.8%) within ±3 standard errors.

Because our normal distribution is centred on the mean height, we can use the properties of the normal distribution to work out how likely we would be to see a difference of at least this size (2.62 inches) if the null hypothesis were true. One way to think about it is as a measure of surprise: if the true difference were zero, how surprised would we be to see a difference of 2.62? To do that we ask where zero lies in the distribution. If zero lies well within the middle of the bell, close to the centre (the observed mean),

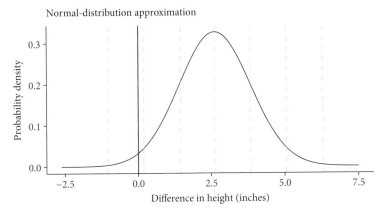

Figure 5.3 Normal-distribution approximation for the differences in plant heights. Dashed lines show ±1, 2, and 3 standard errors around the mean difference in height of 2.62 inches; the red line shows the null hypothesis of zero difference in height.

then we cannot distinguish between the two values: given the level of noise, 2.62 is a plausible mean difference in height to observe even if there were really no difference in height between the groups. However, in this case zero is not in the middle of the bell; it falls in the left-hand tail (the left tail because we have calculated a positive difference in height for the crossed group minus the selfed). The probability distribution is continuous, but if we relate the value of zero to the arbitrary markers (shown by vertical grey lines in the Fig. 5.3) it occurs beyond minus two standard deviations. Because the value of zero lies beyond the zone of the central 95% of the area under the curve, it lies in an extreme part of the distribution (although we should note that it is not far outside the ±2-standard-errors zone). This 'clear blue water' of at least 2 standard errors is conventionally taken as the minimum needed to reject the null hypothesis. We can think of this result as having passed the test but only at the lowest level of confidence. If we increase the severity of our test to require a difference of at least three standard errors then the result fails. In court, juries have to be assured that evidence of a crime is 'beyond reasonable doubt' in order to convict. The convention in statistics has been to see ±2 standard errors as our minimal level of 'beyond reasonable doubt'. Of course, it is important to keep in mind that this means we have to accept that some results guilty of being falsely positive will be unfairly treated, while 'false negatives' are left off the hook.

5.6 Confidence intervals

Because ±2 standard errors covers the central 95% of the area under the normal curve, this is referred to as a 95% confidence interval (or 95% CI). We can calculate intervals corresponding to any level of confidence (±1 standard error is an approximate 68% CI, for example), but 95% intervals are the most commonly used and many statisticians encourage their adoption as the standard thing to report. Confidence intervals are defined by their lower and upper limits (bounds), which we can calculate here as

```
2.62 - 2 * 1.22
```

```
## [1] 0.18
```

```
2.62 + 2 * 1.22
```

```
## [1] 5.06
```

A common mistake is to state that we are 95% confident that the 'true' parameter value (in this case the unknown population mean) is in the interval. But, either the parameter is in the interval (100%) or it is not (0%)—we just don't know. So what does the 95% refer to? Confidence intervals come from the school of *frequentist* statistics, which is named for the idea of a hypothetical repeated series of samples. Confidence intervals are designed so that, for an imaginary long run of repeated samples, the interval will capture the 'true' value in 95% of cases. So, we can say we are 95% confident that our interval includes the 'true' value (in this long-term sense). We will explore these ideas in relation to probability values later.

We could draw a graph of the estimate and interval, but, given that we only have one, it is more efficient to report it as part of our text. The convention is to report the point estimate followed by the upper and lower bounds in square brackets:

> The maize plants that had been cross-pollinated were taller on average than the self-pollinated individuals, with a mean difference in height of 2.62 [0.18, 5.06] inches.

While 95% is the conventional level of confidence to present, intervals often reflect other levels (99% etc.), so there is potential for confusion. It is therefore important to clearly state what is being presented (at least at first use):

> ...2.62 [0.18, 5.06] inches (mean and 95% CI).

The same is true when presenting an estimate ± a margin of error. At the start of this exercise we presented the estimate ±1 standard error. However, estimates are also often presented with a margin of error of ±2 standard errors (i.e. the value to add and subtract from the estimate for a 95% confidence interval). Again, it is better to be explicit (at least at first use):

> ...a mean difference in height of 2.62 (SE: ±1.22) inches.

If you think back to the point about how much we trust the results of polls based on their sample size, it is also important to clearly report n. In this case, we have $n = 15$ paired differences in height. Our approximate 95% CI is ±2 standard errors based on assuming a large sample size. The approximation isn't too bad for $n = 15$ (obviously it depends how precise you want to be) but our result is marginal—close to the boundary of the interval. The result also depends on how well the test meets the underlying assumptions—something we have not assessed. Don't worry; in the next chapter we will repeat this analysis using the more efficient and exact R functions designed for linear-model analysis, including automated methods for assessing the assumptions.

5.7 Summary: Statistics

Statistics is all about signal and noise. When we want to quantify the variability (dispersion) in a sample, the standard deviation is usually the descriptive statistic of first choice. The standard deviation of the sampling distribution of a mean is called the standard error of the mean. The standard error is the basic measure of precision. Thanks to the central limit theorem, we can often use the properties of the normal distribution as a model for the variability in our data, allowing us to form confidence intervals that define the range of plausible values for our estimate. The standard measure of confidence is a 95% confidence interval (which should capture the true value of the estimate 95% of the time).

5.8 Summary: R

- An efficient feature of R is that it does vector-based calculations.
- R has functions to calculate summary statistics, including the sample mean, standard deviation, and variance: mean(), sd(), and var().

Appendix 5a: R code for Fig. 5.1

```
ggplot(data = darwin,
  geom_point(aes(x = type, y = height)) +
  scale_y_continuous(limits = c(10, 25), minor_breaks = seq(10,
    25, 1), breaks = seq(10, 25, 5)) +
  geom_hline(yintercept = c(18.1, 20.5), colour = c("blue", "red"),
    linetype = 2) +
  theme_bw()
```

Linear Models

6.1 Introduction

In the last chapter we conducted a simple analysis of Darwin's maize data using R as a calculator to work out confidence intervals 'by hand'. This is a simple way to learn about analysis and good for demystifying the process, but it is inefficient. Instead, we want to take advantage of the more sophisticated functions that R provides that are designed to perform linear-model analysis. We will explore those functions by repeating and extending the analysis of Darwin's maize data.

6.1.1 R PACKAGES

```
library(arm)
library(DAAG)
library(ggfortify)
library(ggplot2)
library(SMPracticals)
```

The New Statistics with R: An Introduction for Biologists. Second Edition. Andy Hector,
Oxford University Press. © Andy Hector 2021. DOI: 10.1093/oso/9780198798170.003.0006

6.2 A linear-model analysis for comparing groups

R has a general function for fitting linear models: lm(). We began our analysis of Darwin's maize data by estimating the grand mean height of all 30 plants (ignoring the pollination treatments to start with). We can do that using the lm() function as shown below. Usually we want to fit more than one model, and so a common way to work is by creating a named R object for each model so that it is easy to compare them, extract information from them, and so on. For example, here I have called the model ls0, short for least squares model 0, because the linear-model analysis uses a technique called least squares (which we'll explore later in Chapters 7 and 11:

```
ls0 <- lm(formula = height ~ 1, data = darwin)
```

The first argument of the lm() function—the model formula (I've named it here but won't in future)—specifies that we want to analyse the response variable (height) as a function of an explanatory variable, using the tilde symbol, ~. To start with the simplest possible model, we have ignored the pollination-type treatment. Instead, the '1' indicates that we just want to estimate an 'intercept' (we have to have something to the right of the tilde; we can't just leave it blank). When nothing else is included in the linear-model formula apart from the '1', the intercept is the grand mean. Note how the lm() function has an argument for specifying the name of the dataframe (which saves us having to use the with() function or the dataframe$variable notation). The statisticians Andrew Gelman, Jennifer Hill, and colleagues have written a handy display() function as part of their arm package, which gives a concise summary of some the key output of linear models (display() is a simplified alternative to the base-R summary() function, which you can use instead if you don't have the arm package installed):

```
library(arm)
display(ls0)
```

```
## lm(formula = height ~ 1, data = darwin)
##                coef.est coef.se
## (Intercept) 18.88        0.58
## ---
## n = 30, k = 1
## residual sd = 3.18, R-Squared = 0.00
```

The output is called the *table of coefficients*. R first repeats the model formula. Underneath, the headings to the two columns of numbers indicate that 18.88 is the estimate of the model *coefficient* (in this case the grand mean), together with its standard error. The first row of this type of R output is always labelled generically as '(Intercept)' and the challenge is to work out what this intercept is. In this case, we can confirm that the intercept is simply the overall grand mean that we calculated previously using the mean() function:

```
mean(darwin$height)
```

```
## [1] 18.88333
```

The sample size, n, gives the number of data points (the number of rows in the dataframe), and k gives the number of parameters estimated by the linear model, in this case just the grand mean. *R-squared* is the proportion of the variation accounted for by the explanatory variable (the signal). In this model, we have no explanatory variable and so we explain none of the variation. The residual standard deviation of 3.18 is the standard deviation of the heights of all 30 plants that we calculated previously:

```
sd(darwin$height)
```

```
## [1] 3.180953
```

This simple model is a good place to start in order to demystify what the lm() function is doing, but what we really want is a linear model that analyses the difference in average plant height as a function of pollination type. As we explore how to do that, let's also take another closer look at the data. So far, we have failed to note one important property of the data—have you spotted this deliberate omission already, or can you spot it now?

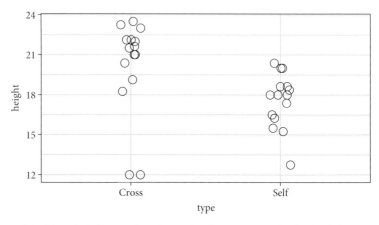

Figure 6.1 Plant height as a function of pollination type. Individual data points are jittered to avoid overplotting.

To avoid distraction, the R code for the graph in Fig. 6.1 is given at the end of the chapter. There are two modifications to the plot. First, the plant heights are shown by open symbols. Second, the data have been horizontally 'jittered' (a small amount of random noise has been added to the data as they are plotted). These are two ways to address overplotting. It turns out that there are two outlying cross-pollinated plants with an unusually low height of 12 inches—something that was not clear from our earlier figures. We can use the lm() function to fit the linear model as follows:

```
ls1 <- lm(height ~ 1 + type, data = darwin)
```

The model formula now contains the pollination type in addition to the intercept indicated by the 1. Actually, we can omit the 1 since R will automatically (but invisibly) include it for us (if you compare this model with one that omits the 1, you will see the output is identical—try it). This linear-model analysis makes some assumptions that we will check later. For now, let's compare the outputs for this model and the last one. Notice that the intercept in model ls0 is no longer the grand mean, as we can see from the display() function output:

```
display(ls1)
```

```
## lm(formula = height ~ 1 + type, data = darwin)
##              coef.est coef.se
## (Intercept) 20.19      0.76
## typeSelf    -2.62      1.07
## ---
## n = 30, k = 2
## residual sd = 2.94, R-Squared = 0.18
```

Now there are two rows of numbers in the output, so what is the intercept in this case? We can work out what the intercept is in the style of Sherlock Holmes—by eliminating all of the possibilities bar one. The label of the second row, typeSelf, is produced by combining the name of the explanatory variable (type) followed by 'Self'—the name of one of the levels of this factor. Since type has only two levels, the intercept must be the other one, and could be relabelled as 'typeCross'. So, the coefficient in the row labelled 'Intercept' is the average height of the 15 maize plants in the crossed treatments. Notice that the labels of the rows combine the names of the factor and the factor level with no spacing. In this case, the use of a leading capital letter for the factor level helps break things up (another trick when making your own dataframes is to end the name of the factor with an underscore, so that then the row would be type_Self). A common mistake is to think that the value in the second row is the height of the selfed plants. This would seem a fair assumption, as the label 'typeSelf' implies this is what it ought to be. However, in this case it is obvious that this can't be the mean height of the selfed plants, since the value is negative! Instead, the output shows the *difference* between means. So, in this case the intercept gives the mean height of the crossed plants and the second row gives the difference in the mean heights of the two groups. This may seem a bit odd at first sight, but there are good reasons for doing it this way, and in this case it also focuses our attention on the question of interest: is there any difference in plant height that results from whether pollination was through selfing or outcrossing? Note that the output no longer gives the overall grand mean: our question does not involve the grand mean and estimating it would cost us an extra, wasteful parameter. Finally, adding the pollination-type treatment to the model explains 18% of the variation in plant height. See Box 6.1.

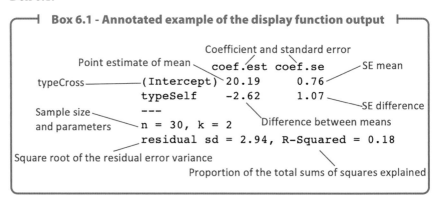

Box 6.1 - Annotated example of the display function output

```
                             Coefficient and standard error
        Point estimate of mean    coef.est coef.se        SE mean
  typeCross           (Intercept) 20.19       0.76
                      typeSelf     -2.62       1.07
                                                          SE difference
  Sample size         ---
  and parameters      n = 30, k = 2      Difference between means
                      residual sd = 2.94, R-Squared = 0.18
  Square root of the residual error variance
                              Proportion of the total sums of squares explained
```

Note that the output for model ls1 has not given us the mean height for the selfed plants. We can calculate the mean of the other group by subtraction, but we can't do the same for the standard error. At the end of the chapter we'll see how to get R to 'relevel' to make the selfed treatment the intercept instead. However, for now we can produce a graph of the individual data with the means superimposed. First, Fig. 6.2 shows a quick plot that makes the self- and cross-pollinated plants more distinct using different colours for the two treatment groups (we can use the shape argument in the same way to alter the shape too):

```
fig_6.2 <- qplot(data = darwin, x = type, y = height, colour = type)
fig_6.2
```

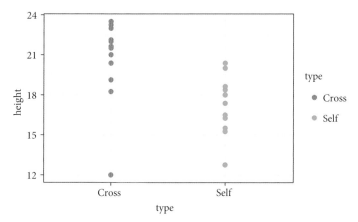

Figure 6.2 Plant height as a function of pollination type (indicated by colour).

We can now superimpose the means (using black stars to stand out from the coloured points—the R code is shown at the end of the chapter) (Fig. 6.3):

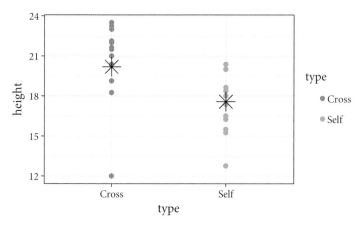

Figure 6.3 Plant height as a function of pollination type (black stars show treatment means).

6.3 Standard error of the difference

The second column of the output of the display() (or summary) function gives the standard errors of the values in the first column. Since the first column of numbers contains a mean and a difference between means (in the first and second rows), the second column gives a standard error of the mean (SEM) and a standard error of the difference between the two means (SED). We have already met the SEM in the last chapter, but what is the SED? The formula for the standard error of the difference between two means is

$$SED = \sqrt{\frac{s_1^2}{n_1} + \frac{s_2^2}{n_2}}$$

where the subscripts 1 and 2 indicate the two groups—the pollination treatments in this case. When the standard errors for the two groups have the same variance and the same sample size, then the SED is about 1.4 (the square root of 2) times larger than the SEM:

$$SED = \sqrt{2} * SEM$$

In the last chapter we calculated the variances and standard deviations for the cross- and self-pollinated plants and found that they were different. However, the linear-model analysis takes a different approach. Rather than calculate separate variances for each group, the linear model calculates a more accurate 'pooled' variance across all groups to take advantage of the full sample. Of course, this does assume that the groups share the same variance, an assumption we must check in linear-model analysis.

6.4 Confidence intervals

R has a confint() function for calculating confidence intervals. The confint() function output follows the same layout as the table of coefficients that gives the point estimates:

```
display(ls1)
```

```
## lm(formula = height ~ 1 + type, data = darwin)
##               coef.est coef.se
## (Intercept)   20.19    0.76
## typeSelf      -2.62     1.07
```

```
## ---
## n = 30, k = 2
## residual sd = 2.94, R-Squared = 0.18
```

Remember that the display() output shows the mean height for the crossed maize plants (20.19 inches) and the difference in mean height between the crossed and selfed plants (−2.62 inches), both with their respective standard errors. In the earlier chapter we worked with the positive difference in height of the crossed plants relative to the selfed ones, but here the sign is reversed.

The confint() function will calculate the lower (2.5%) and upper (97.5%) bounds of a 95% CI based on these standard errors:

```
confint(ls1)
```

```
##                    2.5 %      97.5 %
## (Intercept) 18.63651 21.7468231
## typeSelf     -4.81599 -0.4173433
```

Because it follows the same layout as the table of coefficients, the confint() output 'intercept' row gives a 95% CI for the height of the crossed plants and the second row gives a 95% interval for the difference in height between the crossed and selfed plants. The lower and upper bounds are the 2.5 and 97.5 percentiles of the normal distribution. It may be helpful to bring the output of the various functions together in one table (Table 6.1):

It is the difference in height that addresses the question that motivated Darwin's experiment. Specifically, Darwin hypothesized that self-pollinated plants would have lower fitness (as reflected in lower height)

Table 6.1 Coefficients with standard errors and 95% confidence intervals.

	Estimate	Standard error	Confidence interval (95%)
Crossed	20.2	0.76	18.64, 21.75
Difference	−2.6	1.07	−4.82, −0.42

than cross-pollinated plants. We can use the confidence intervals to answer Darwin's question.

6.5 Answering Darwin's question

Darwin's hypothesis was that self-pollination would reduce fitness (using height as a surrogate). The null hypothesis is of no effect of pollination type and therefore no difference in height. Is this null hypothesis consistent with the results of the experiment or can we reject it? It is simple to use the confidence interval to test the null hypothesis. To do so, we simply see whether or not the predicted null value lies inside the confidence interval. If it does, we cannot distinguish the observed mean difference in height (−2.62 inches) from the null prediction of zero difference in height, given the level of variability (noise) in the data. The variability is quantified by the standard error that forms the basis for calculating the CI. In this case we can see that both the lower and the upper bound of the confidence interval are negative numbers—zero lies outside the confidence interval and we can reject the null hypothesis. The difference in height is consistent with Darwin's hypothesis of inbreeding depression.

The arm package has a coefplot() function that produces a graph of a point estimate (our estimate of the difference in height, −2.62 inches) with an approximate 68% CI (±1 SE) and 95% CI (±2 SEs) (Fig. 6.4). The null hypothesis value of zero is outside of the confidence interval, it is not consistent with the data, and we can reject it at this level of confidence (the xlim argument is needed here to extend the x-axis limit to show zero in the figure):

```
library(arm)
fig6_4 <- coefplot(ls1, xlim = c(-5, 0))
```

Following R.A. Fisher, we either reject the null hypothesis or fail to reject it. And, of course, if we increase the level of confidence (say from 2 standard errors to 3) then a result that achieved a lower standard may fail to clear the higher bar. For example, we could produce a 99% CI:

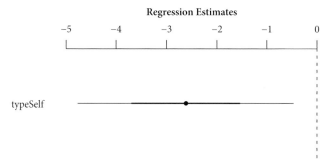

Figure 6.4 A coefplot() graph showing the difference in height with intervals of ±1 and ±2 SEMs.

```
confint(ls1, level = 0.99)
```

```
##                     0.5 %      99.5 %
## (Intercept)  18.093790 22.2895433
## typeSelf      -5.583512  0.3501789
```

Now, the change in sign in the second row shows that zero now lies within the interval—we cannot reject the null hypothesis at this higher level of confidence. The coefplot() function works in numbers of standard errors (or sds, as it refers to them)—setting it to show ±3 standard errors (Fig. 6.5) is similar to using the 99% CI with the confint() function and produces the same qualitative result, where zero lies within the interval:

```
fig6_5 <- coefplot(ls1, sd = 3)
```

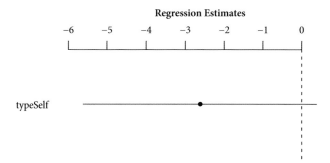

Figure 6.5 A coefplot() graph showing the difference in height with an interval of ±3 SEMs.

Overall, if we view the different levels of confidence intervals as tests of increasing severity then the reduction in height of the self-pollinated plants is large enough relative to the background noise that it supports Darwin's hypothesis at the lower level of confidence, but not at the higher level.

6.6 Relevelling to get the other treatment mean and standard error

One limitation of the table-of-coefficients output given above is that we don't have the mean and standard error of the other treatment level. It is easy enough to get the mean by addition or subtraction, but the standard error cannot be calculated in this way. Instead, we can get R to estimate the mean and standard error for the selfed plants by using the relevel() function to set this treatment as the intercept:

```
darwin$type <- relevel(darwin$type, ref = "Self")
display(lm(height ~ type, data = darwin))
```

```
## lm(formula = height ~ type, data = darwin)
##                coef.est coef.se
## (Intercept) 17.58      0.76
## typeCross    2.62      1.07
## ---
## n = 30, k = 2
## residual sd = 2.94, R-Squared = 0.18
```

After relevelling, the selfed treatment level is taken as the intercept, and now we get the point estimate for its mean and standard error of the mean in the top row and the estimated difference in means (same value, opposite sign) and SED in the row below. Notice that the SEM for the selfed treatment is the same as that for the crossed treatment. If we look at the top and bottom of the formula for the standard error,

$$SE = \sqrt{\frac{s^2}{n}}$$

we can see that SEMs for different treatments can be different if the variance or the sample size (or both) varies among groups. The variance can differ if SEMs are being calculated 'by hand' (usually in a spreadsheet) and the variances are estimated separately for each group. However, here the standard errors are being provided by the linear model-function, which uses a single (pooled) estimate of the residual variance for all treatment levels (we'll come back to this when we meet the residual variance in ANOVA tables). So, if the SEMs presented by a linear-model analysis vary, we know this must be due to differences in the sample sizes of the different treatments. In this case we know that both treatments had the same number of plants (15, with no missing values). In fact, because we also knew that the lm() function uses a pooled estimate of the variance, we did not need to relevel to know that the standard error of the mean for the selfed plants was the same as the SEM for the crossed maize—but, again, it is good to work through this once to help demystify what the lm() function is doing.

6.7 Assumption checking

We have now been through the key parts of the results of the linear-model analysis, but before we can trust them we need to check that the assumptions of the model are adequately met. Here we are going to focus on the assumptions that the unexplained variations in the two treatment groups are approximately normal and are equal in their variability. Statisticians usually recommend doing this graphically using plots of the residual differences. As the name implies, residuals are the differences between the observed values (heights) and the 'fitted' values predicted by the linear model—in this case the treatment means (see Chapter 7). The assumption of approximate normality applies because linear models use the normal distribution as a model for the variability, from which we can derive measures of precision (standard errors) and confidence (confidence intervals). The assumption of approximately equal ('homogeneous') variability follows from the use of the single pooled estimate of the variance across all treatment groups (here we have two treatments with 15 replicates each, but often there are more treatments with smaller sample sizes for each). We will look at assumption checking in greater depth later, but for now we can make an initial simple application of these checks. Luckily, functions exist in R that automatically produce some of the most common residual-checking graphs. The ggfortify package (a companion to ggplot2) has an autoplot() function. The same graphs can be produced with the base version of R using the generic plot() function when applied to a linear-model object (in this case, plot(ls1)).

6.7.1 NORMALITY

Approximate normality can be judged using a 'normal quantile' plot that graphs the quantiles of the distribution of the residuals versus the quantiles of a random sample of normally distributed numbers so that the residuals fall around a straight line when they match the model:

```
library(ggfortify)
fig6_6 <- autoplot(ls1, which = c(2), ncol = 1)
fig6_6
```

In this case (Fig. 6.6) we can see that a large portion of the residuals are approximately normally distributed except at the extremes—this is not unsurprising, given that we have already noted outlying values in our initial exploration and description of the data. We will look at non-normality in greater depth later, but for now we can conclude that while this case is not terrible, it is far from perfect. Luckily, linear-model analysis is fairly robust to most forms of non-normality (and 'robust' extensions to the linear-model function exist, such as the rlm() function in the MASS package—see below).

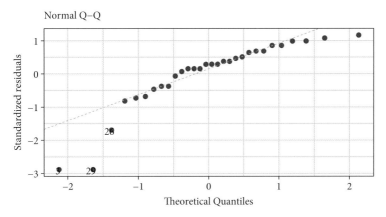

Figure 6.6 Normal quantile–quantile plot used to assess normality of the residuals.

6.7.2 EQUAL VARIANCE

We have two graphs available to us from the automatic model-checking plots that are most useful for assessing if the variances are equal. They both plot the residuals versus the fitted (predicted) values—a residual of zero means an observation was bang on the predicted value—the treatment mean in this case. We can look at the raw residuals (the left-hand panel in Fig. 6.7), which are positive and negative, or as standardized absolute values—where we divide the residuals by their standard deviation, make all the values positive and take their square root (right panel):

```
fig6_7 <- autoplot(ls1, which = c(1, 3), ncol = 2)
fig6_7
```

In this case both plots confirm our impressions from graphs of the raw data—the group with the higher fitted values (the cross-pollinated treatment with the higher mean) is a bit more variable, mainly due to the outlying values we have also already noted. Once again, this is not terrible, but not perfect. It is particularly tricky, as the greater variation is generated

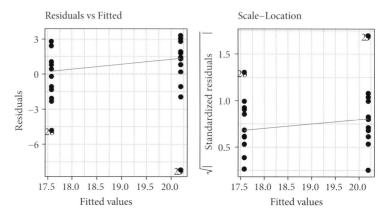

Figure 6.7 Diagnostic plots used to assess whether residuals show approximately constant variance.

by just a few outliers. There are three more residual plots available that, among other things, would allow us to investigate the influence of the outlying values in greater detail, but we are going to leave that for later.

In this chapter we started with a simple linear-model analysis of Darwin's maize data. However, we have not addressed the fact that the maize plants come in pairs of crossed and selfed plants, nor considered which pot the pairs of plants were planted in. The main reason for this is purely to keep things simple. However, as we'll see later in Chapter 11, the pairing was not done properly and consequently has not worked as intended, so it is debatable whether it is best to analyse this as a paired design or not.

6.8 Summary: Statistics

Linear-model analysis usually sets one factor level as the 'intercept', estimates its mean, and then gives the differences between this intercept and the means of the other factor levels. The differences in means are accompanied by standard errors—SEDs. In simple situations the SED is approximately 1.4 times the SEM (SED = SEM × sqrt(2)). Standard errors can be used to calculate confidence intervals—so long as sample sizes are not very small, an approximate 95% CI is ±2 standard errors. Intervals can be calculated to reflect differing degrees of confidence—95% confidence intervals are the most commonly presented (although they are the lowest level of confidence conventionally used). Intervals can be used to perform statistical tests (although they can do a lot more than that). Statistical tests usually attempt to reject a 'no effect' null hypothesis (or fail to reject it). Linear-model analysis makes various assumptions, including that the noise (residual differences) is approximately normally distributed with roughly equal ('homogeneous') variance.

6.9 Summary: R

- Linear-model analysis is conducted in R using the lm() function.
- The lm() function default is to use alphanumeric rules to choose which factor level is the 'intercept'.
- The relevel() function is used to override the alphanumeric default and set which factor level is used as the intercept.
- Confidence intervals for linear-model coefficients are calculated by the confint() function.

6.10 Reference

Maindonald, J. & Braun, J.W. (2010) *Data Analysis and Graphics*. Cambridge University Press.

Appendix 6a: R graphics

Darwin's maize data can be jittered as in Fig. 6.1 as follows:

```
ggplot(data = darwin, aes(x = type, y = height)) +
geom_jitter(shape = 1, size = 3, width = 0.07, height = 0)
```

The treatment means can be superimposed as in Fig. 6.3 as follows:

```
qplot(data = darwin, x = type, y = height, colour = type) +
   stat_summary(fun = mean, geom = "point", colour = "black",
      shape = 8, size = 5)
```

Appendix 6b: Robust linear models

The MASS package has a robust linear-model function, rlm(). Note the change in the estimates and standard errors. We'll explore *t*-tests in a later chapter dedicated to that topic.

```
library(MASS)
summary(rlm(height ~ type, data = darwin))
```

```
##
## Call: rlm(formula = height ~ type, data = darwin)
## Residuals:
##     Min      1Q  Median      3Q      Max
## -9.1542 -1.4374  0.2501  0.9469  2.6251
##
## Coefficients:
##               Value   Std. Error t value
## (Intercept) 17.7499  0.6362      27.8989
## typeCross    3.4043  0.8998       3.7836
##
## Residual standard error: 1.646 on 28 degrees of freedom
```

Appendix 6c: Exercise

Darwin's 1876 book reports many experiments, including one on wild mignonette which has the same general design as his one on maize. Repeat the maize analysis for wild mignonette—does it show the same results?

The mignonette data (in wide format) is available in the DAAG package that accompanies the book *Data Analysis and Graphics* by Maindonald and Braun (2010):

```
library(DAAG)
str(mignonette)
```

Regression

7.1 Introduction

So far we have met linear models for analyses with categorical explanatory variables ('factors'), often known as analysis of variance (ANOVA). But what if your explanatory variable is continuous, not categorical? For that you need a closely related linear model known as regression. The good news is that ANOVA and regression are so close that R performs them with the same linear-model function, lm(), meaning you do not need to learn any new R functions, only how to interpret the output in this new context.

7.1.1 R PACKAGES

```
library(arm)
library(ggplot2)
library(MASS)
library(SemiPar)
```

The New Statistics with R: An Introduction for Biologists. Second Edition. Andy Hector,
Oxford University Press. © Andy Hector 2021. DOI: 10.1093/oso/9780198798170.003.0007

7.2 Linear regression

Once again, we can think of linear regression analysis in terms of signal and noise. As the name implies, we use linear regression when we think the relationship between a response and an explanatory variable is approximately straight (although, as we'll see later, it is possible to modify it to fit curvilinear relationships). Regression analyses are usually visualized with scatter plots of the continuous response as a function of the continuous explanatory variable. Essentially, we want a straight-line relationship that goes through the middle of the cloud of data points (see Fig. 7.1) to 'model the mean'. The variability around that 'mean' is captured by the size of the standard errors and is reflected in the confidence intervals calculated from them. The linear regression model can be written as

$$y = a + bx$$

where

- y is the predicted value of the response variable (hardness in the example below);
- a is the regression intercept (the value of y for $x = 0$);
- b is the slope of the regression line;
- x is the value of the explanatory variable (density—dens—in the example below).

This formulation omits the unexplained residual (error) variation, which can be added as an additional term, e:

$$y = a + bx + e$$

The regression uses two values to define a straight line. First, we need a starting point for the line: the value of the so-called regression intercept. The regression intercept is the value of y when $x = 0$. Then we need a gradient (how the value of y changes as the value of x changes) that allows us to draw the line from that point: the value of the regression slope. The linear-model analysis estimates the values of the intercept and slope (along with standard errors). In the example below, we will use the estimated regression intercept and slope to predict the hardness of a timber sample with a particular wood density.

7.3 The Janka timber hardness data

Our regression analysis uses example data from the Australian forestry industry that record the density and hardness of 36 samples of wood—presumably from 36 different tree species (although the names are not given). Density is a fundamental property that is easier to measure than wood hardness—which is measured here by recording the force (in 'pounds-force') required to push a ball bearing into a timber sample using a mechanical press. In contrast to Darwin's maize example, the goal of this exercise is not to test a biological hypothesis. Instead, the aim is to establish a linear regression relationship between wood density (our explanatory variable, or predictor) and timber hardness (the response). We can then use the estimated regression line to predict timber hardness for samples with known density but unknown hardness. Timber hardness has been quantified using the Janka scale. The data originally come from a 1959 book, and the regression analysis was developed by William (Bill) Venables (2000) with input from fellow statistician John Nelder. The Janka data can be obtained from the SemiPar package (often, once a package is loaded using the library() function the data are immediately available by name, but for packages like SemiPar there is an extra step where the data are loaded using the data() function):

```
library(SemiPar)
data(janka)
head(janka)
```

```
##    dens hardness
## 1 24.7      484
## 2 24.8      427
## 3 27.3      413
## 4 28.4      517
## 5 28.4      549
## 6 29.0      648
```

The output of the str() function tells us that the janka data set is 36 rows long and has two columns. For each of 36 samples, we get the wood density ('dens') and timber hardness values explained above:

```
str(janka)
```

```
## 'data.frame':    36 obs. of  2 variables:
## $ dens    : num  24.7 24.8 27.3 28.4 28.4 29 30.3 32.7 35.6 38.5 ...
## $ hardness: int   484 427 413 517 549 648 587 704 979 914 ...
```

The density has (presumably) been given a short name to avoid duplicating the R function of the same name (by convention, variable names are all lower case—remember R is case sensitive). A plot of hardness (the response) as a function of density (the explanatory variable) looks like Fig. 7.1:

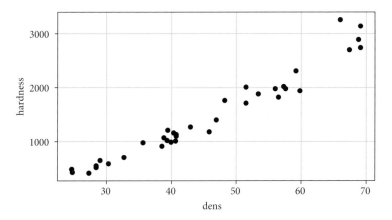

Figure 7.1 Timber hardness as a function of wood density.

```
fig7_1 <- qplot(data = janka, x = dens, y = hardness) +
    theme_bw()
fig7_1
```

7.4 Correlation

Wood density and timber hardness appear to be positively related and the relationship seems to be fairly linear. We can look at the strength of the association between dens and hardness using correlation (the cor() function has no data argument, so we need to use with()):

```
with(janka, cor(dens, hardness))
```

```
## [1] 0.9743345
```

The correlation coefficient ranges from -1 for perfectly negative relationships through zero for unrelated variables to $+1$ for perfectly positively related variables. The relationship here is strongly positive (the related cor.test() function will perform a test, but that is not our primary interest here). Correlation looks at the association between variables, but here we can go further—we are arguing that higher wood density causes higher values of timber hardness. When we are willing to assume that changes in x lead to changes in y, we can go further than correlation and use linear regression.

7.5 Linear regression in R

We can fit the linear regression model in R in the same way as for Darwin's cross- and self-pollinated maize plants—the only difference is that our explanatory variable, density, is continuous rather then being categorical (be careful not to reverse the order of the variables—here we want to predict hardness from the more fundamental property of density):

```
janka.ls1 <- lm(hardness ~ dens, data = janka)
```

The linear-model analysis estimates a 'line of best fit' (the regression line) by using the method of least squares to minimize the error sums of squares (the average distance between data points and the line, i.e. the unexplained variation; see Appendix 7b). We can add the regression line to the scatter plot as follows (Fig. 7.2):

```
fig7_2 <- fig7_1 + geom_smooth(method = "lm")
fig7_2
```

The new line of R code adds a 'smoother'—a line—to the graph, in this case a straight linear regression line from a linear model (method = "lm") along with a 95% confidence interval band. Note that the confidence interval for the regression line has curved upper and lower bounds that are narrowest in the middle. This is because it is calculated using the standard errors for the regression intercept and the regression slope. There is therefore uncertainty about both the elevation of the line (how high up the graph it is) and its gradient (how steep or shallow the slope is).

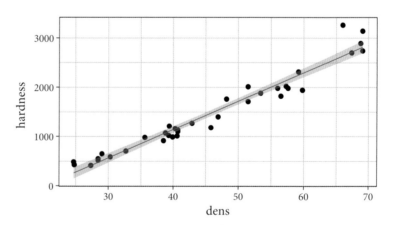

Figure 7.2 Linear regression of timber hardness as a function of wood density with 95% CI.

If you imagine moving the line up and down while wiggling it to make the slope steeper or shallower, you can understand the shape of the confidence interval.

At this point during a real analysis, we would usually use the residual-checking plots to assess how well our linear model has met its assumptions before looking at the results. However, for our purposes we'll postpone that until a little later. The first thing to look at is the estimates of the coefficients, which we can do using the arm package's display() function (if you don't have the arm package, use the base-R summary() function instead):

```
library(arm)
display(janka.ls1)
```

```
## lm(formula = hardness ~ dens, data = janka)
##                 coef.est coef.se
## (Intercept) -1160.50     108.58
## dens            57.51       2.28
## ---
## n = 36, k = 2
## residual sd = 183.06, R-Squared = 0.95
```

This output looks very similar to that produced for the analysis of Darwin's maize data, including a column of coefficient estimates (coef.est) and a column of their standard errors (coef.se). We just have to interpret what these estimates are, now that we have a continuous explanatory variable (density) instead of a factor. In some ways, it is easier to interpret the display() and summary() function output for regression models, because the generic label in the first row, (Intercept), is more appropriate here, since this row reports the regression intercept as defined above. So, -1160 is the estimated value of y (hardness) when x (density) $= 0$. The value to its right is simply the standard error of the regression intercept. Obviously, it is not possible to have a negative hardness value. We get this impossible hardness value because the regression intercept is the value of y when $x = 0$ and in this case that is a wood density value of zero—also impossible.

Because of how the regression intercept is defined, sometimes the values will be meaningful and sometimes not (as here)—it is up to us as analysts to work this out. One solution is the technique of centering, which subtracts the average x-value (density) from each individual value—imagine moving the vertical y-axis to the right from a value of $x = 0$ to the average value of x. Centering is a useful technique for making regression intercepts more meaningful, since they then give the value of the response for the average value of the explanatory variable (we won't use it here, but will come back to it later in the generalized-linear-model analysis of binary data).

The second row is labelled 'dens'. Density is our explanatory variable, which is on the x-axis of our scatter plot (Fig. 7.2), and the slope is estimated against it. So, 57.51 is the value of the regression slope (with its standard error)—that is, the change in the value of y (hardness) for every unit change in x (density). So, an increase in wood density by 1 unit increases the timber hardness by 57.51 units on the Janka scale. The sample size is given as n, and the number of estimated parameters (regression intercept and slope) as k. The last number in the output gives the R-squared: the proportion of the variation in the data explained by the linear regression analysis.

As in the analysis of Darwin's data, the upper and lower bounds of confidence intervals (a 95% CI by default) can be extracted using the confint() function:

```
confint(janka.ls1)
```

```
##                      2.5 %       97.5 %
## (Intercept)  -1381.16001  -939.83940
## dens             52.87614    62.13721
```

7.6 Assumptions

Our regression model makes the same assumptions as all linear models. These include the assumptions that the unexplained variability around the regression line—the residuals—is approximately normal and has constant

variance. We can check these assumptions using the same graphical methods that we used in the analysis of Darwin's maize data. The residuals are
the differences between the observed data and the 'fitted values' predicted
by the model. In other words, the residual differences are the vertical
differences between the data points and the corresponding points on the
regression line (the fitted values). The graph in Fig. 7.3 uses red lines to
indicate one positive and one negative residual (see Appendix 7b):

```
fig7_3 <-
  fig7_1 + geom_smooth(method = "lm", se = FALSE) +
  geom_segment(aes(x = 66, xend = 66, y = 3250, yend = 2650),
    colour = "red") +
  geom_segment(aes(x = 59.8, xend = 59.8, y = 1955, yend = 2265),
    colour = "red")
fig7_3
```

Each observed value of the response has a corresponding 'fitted value'
predicted by the linear model—the red lines indicate the points of the fitted
values on the regression line for the two highlighted data points. The fitted
values can be extracted using the fitted() function:

```
fitted(janka.ls1)
```

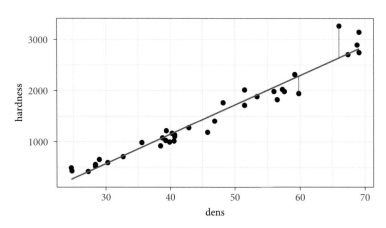

Figure 7.3 Linear regression showing one positive and one negative residual
difference (red lines).

```
##          1         2         3         4         5
##   259.9152  265.6658  409.4325  472.6899  472.6899
##          6         7         8         9        10
##   507.1939  581.9525  719.9686  886.7379 1053.5073
##         11        12        13        14        15
## 1070.7593 1099.5126 1105.2633 1134.0166 1157.0193
##         16        17        18        19        20
## 1174.2713 1180.0220 1180.0220 1306.5366 1473.3060
##         21        22        23        24        25
## 1536.5633 1611.3220 1801.0940 1801.0940 1910.3567
##         26        27        28        29        30
## 2059.8741 2088.6274 2134.6328 2151.8848 2243.8954
##         31        32        33        34        35
## 2278.3994 2634.9408 2715.4502 2795.9595 2813.2115
##         36
## 2813.2115
```

The residuals can be extracted using the residuals() function (nesting it inside the head() and tail() functions shows only the first six and last six values to save space):

```
head(residuals(janka.ls1))
```

```
##          1         2         3         4         5
## 224.084837 161.334170   3.567483  44.310140  76.310140
##          6
## 140.806135
```

```
tail(residuals(janka.ls1))
```

```
##         31        32        33        34        35
## -338.39945  625.05917  -15.45018  94.04048  -73.21152
##         36
##  326.78848
```

Now that we know what residuals are, we can apply the automatic model-checking residual graphs provided by the ggfortify package (if the ggfortify

package is not loaded, applying the base-R plot() function to the linear
model will produce similar plots). First, normality (Fig 7.4):

```
library(ggfortify)
fig7_4 <- autoplot(janka.ls1, which = c(2), ncol = 1)
# plot(janka.ls1, which = c(2))
fig7_4
```

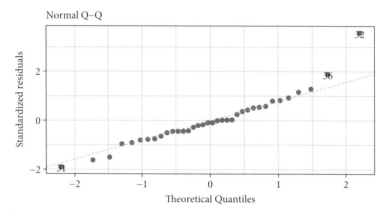

Figure 7.4 Assessing the assumption of normality for the Janka timber hardness
data using a normal quantile–quantile diagnostic plot.

The residuals from the analysis of the Janka data are generally fairly
normally distributed, but, as is often the case, we find some outlying values
(individually numbered on the plot). Now the variability (Fig. 7.5):

```
fig7_5 <- autoplot(janka.ls1, which = c(1, 3), ncol = 2)
fig7_5
```

The left-hand graph plots the raw residuals as a function of the fitted
values. The right-hand 'scale–location' plot does a few things to the residuals. First, it takes the absolute values to increase the density of the data

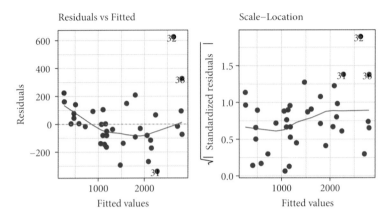

Figure 7.5 Assessing the assumption of approximately constant variance for the Janka timber hardness data using a Tukey–Anscombe plot of residuals versus fitted values (left) and a scale–location plot (right).

points (imagine folding the lower, negative part of the left-hand plot up and overlaying it on the upper, positive section). It also standardizes the residuals by dividing each value by their standard deviation, and finally takes the square root. Both plots suggest that the residuals do not have constant variance—instead, their variability seems to increase with the mean. There is more going on than just the increasing residuals, and we will examine them in greater depth later on as we explore whether it is possible to find a better way to model these data.

7.7 Summary: Statistics

Linear-model analysis with continuous explanatory variables is called linear regression. Linear regression models the relationship between response and explanatory variables using straight lines defined by a regression intercept and slope. The regression intercept is the value of y when $x = 0$. Linear regression models assume the unexplained variability around the line is approximately normal with constant variance.

7.8 Summary: R

- The correlation—the strength of association—between two variables can be quantified using the cor() function.
- Linear regression models can be fitted with the lm() function.
- Straight and curved lines ('smoothers') can be added to ggplot2 graphs using the geom_smooth() function.

7.9 Reference

Venables, W.N. (2000) Exegeses on linear models. *S-PLUS User's Conference 1998.*

Appendix 7a: R graphics

We often end up drawing the same graph in multiple versions. A useful time-saver is to create R objects to define the axis labels. Then, if we decide to change the label, we need do so in one place only rather than in the code for every version of the graph. First, we can make x and y labels (called 'xlabel' and 'ylabel' here)—note the use of the superscript:

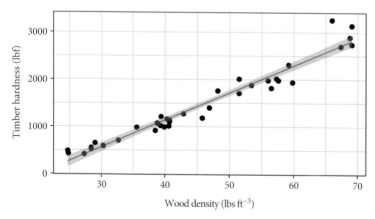

Figure 7.6 Linear regression of timber hardness as a function of wood density with full axis labels.

```
xlabel <- expression(paste("Wood density (lbs ", ft^-3, ")"))
ylabel <- "Timber hardness (lbf)"
```

Then we can substitute them for the *x*- and *y*-variable (column) names using the xlab and ylab arguments (Fig. 7.6):

```
fig7_6 <-
  qplot(data = janka, x = dens, y = hardness,
    xlab = xlabel, ylab = ylabel) +
  geom_smooth(method = "lm")
fig7_6
```

Appendix 7b: Least squares linear regression

The linear-model analysis finds the regression 'line of best fit' using the method of least squares to minimize the average 'residual' distance between the data points and the line (the unexplained variation) (Fig. 7.7). See Chapter 11 for further discussion of least squares:

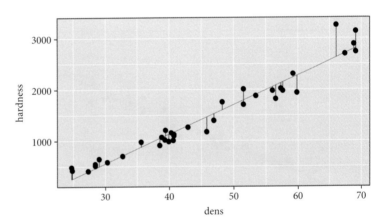

Figure 7.7 The least squares method finds the linear regression 'line of best fit' (grey) to the data (black filled circles) by minimizing the sum of the squared residual differences (vertical red lines).

Prediction

8.1 Introduction

I n the last chapter we used linear regression to model the relationship
between wood density and timber hardness. Our first goal was to
estimate the coefficients of the linear model—the regression intercept and
slope. The aim of this chapter is to put those coefficients to work in
predicting timber hardness from wood density.

8.1.1 R PACKAGES

```
library(dplyr)
library(ggplot2)
library(SemiPar)
```

8.2 Predicting timber hardness from wood density

Our linear regression model from the last chapter (janka.ls1) estimates the
relationship between wood density and timber hardness based on 36 wood
samples for which the density and hardness are both known, ranging in

The New Statistics with R: An Introduction for Biologists. Second Edition. Andy Hector,
Oxford University Press. © Andy Hector 2021. DOI: 10.1093/oso/9780198798170.003.0008

density from 20 to 70 pounds per cubic foot and in hardness up to around 3000 units on the Janka scale:

```
library(SemiPar)
data(janka)
janka.ls1 <- lm(hardness ~ dens, data = janka)
```

Now imagine we have a new wood sample of known density—let's say with a value of 65—but unknown hardness. We can predict what the hardness would be by plugging the values of the estimated coefficients into the equation for the linear regression,

$$y = a + bx$$

The estimates of the intercept and slope are

```
coef(janka.ls1)
```

```
## (Intercept)          dens
## -1160.49970     57.50667
```

which gives us

```
-1160.4997 + 57.50667 * 65
```

```
## [1] 2577.434
```

Rather than work to the number of decimal places given in the output, and with the risk of typing errors, we can instead extract the coefficients from the linear model using square brackets that index the first (intercept) and second (slope) coefficients:

```
coef(janka.ls1)[1] + coef(janka.ls1)[2] * 65
```

```
## (Intercept)
##    2577.434
```

We can also use the predict() function:

```
predict(object = janka.ls1, newdata = list(dens = 65))
```

```
##        1
## 2577.434
```

The first argument is the object—the model—to be used to make the prediction and the second argument lists the values (the 'new data') we want to make a prediction for. By default, a prediction is made for each value of the predictor present in the data, in this case the 36 density values:

```
predict(object = janka.ls1)
```

```
##         1          2          3          4          5
##   259.9152   265.6658   409.4325   472.6899   472.6899
##         6          7          8          9         10
##   507.1939   581.9525   719.9686   886.7379 1053.5073
##        11         12         13         14         15
## 1070.7593 1099.5126 1105.2633 1134.0166 1157.0193
##        16         17         18         19         20
## 1174.2713 1180.0220 1180.0220 1306.5366 1473.3060
##        21         22         23         24         25
## 1536.5633 1611.3220 1801.0940 1801.0940 1910.3567
##        26         27         28         29         30
## 2059.8741 2088.6274 2134.6328 2151.8848 2243.8954
##        31         32         33         34         35
## 2278.3994 2634.9408 2715.4502 2795.9595 2813.2115
##        36
## 2813.2115
```

These are the 'fitted values' that we met at the end of the last chapter when introducing residuals—the points on the regression line corresponding to each observed data point. R calculates the fitted values (predictions) by combining the estimates of the coefficients with the so-called model matrix (or X-matrix). One nice feature of R is that it is easy to see the model matrix using the function of the same name. To save space, we can just look at the first few rows,

```
head(model.matrix(janka.ls1))
##   (Intercept) dens
## 1           1 24.7
## 2           1 24.8
## 3           1 27.3
## 4           1 28.4
## 5           1 28.4
## 6           1 29.0
```

and the last few,

```
tail(model.matrix(janka.ls1))
##    (Intercept) dens
## 31           1 59.8
## 32           1 66.0
## 33           1 67.4
## 34           1 68.8
## 35           1 69.1
## 36           1 69.1
```

Understanding the model matrix helps our understanding not only of linear models but also of their extensions, including mixed-effects models (which combine the model matrix with a second matrix). Note that the first column in the model matrix is headed '(Intercept)'—the same label we are already familiar with from the summary() and display() function outputs. This column is always a vector of ones. As we saw when introducing the linear-model function, the model formula always starts with a '1', which indicates the intercept, and because this is always the case R invisibly includes the '1' so that you rarely see model formulas written out to explicitly include it, as we can do here as follows:

```
lm(formula = 1 + hardness ~ dens, data = janka)
```

The column of ones indicates that the value of the intercept is included as the starting point for the calculation (multiplying the value of the intercept by one), which then adds the products of the values of the density and the value of the slope:

$$y = a * 1 + b * x$$

For example, to calculate the predicted hardness for the wood sample with the lowest density, we take the estimated values of the intercept and slope,

```
coef(janka.ls1)
```

```
## (Intercept)          dens
## -1160.49970      57.50667
```

and combine them with the values from the first row of the model matrix,

```
model.matrix(janka.ls1)[1, ]
```

```
## (Intercept)          dens
##          1.0          24.7
```

as follows (with a little rounding error):

```
-1160.4997 * 1 + 57.50667 * 24.7
```

```
## [1] 259.915
```

Or, as R does it (more precisely),

```
coef(janka.ls1)[1] * model.matrix(janka.ls1)[1, 1] +
coef(janka.ls1)[2] * model.matrix(janka.ls1)[1, 2]
```

```
## (Intercept)
##    259.9152
```

8.3 Confidence intervals and prediction intervals

One of the great things about the ggplot2 package is that we can use the qplot() function to easily plot a variety of smoothers (from simple linear regression slopes to more complex curves) together with their confidence intervals. For example, we can easily plot the linear regression with its 95% CI (Fig. 8.1):

```
library(ggplot2)
Fig8_1 <- qplot(data = janka, x = dens, y = hardness) +
  geom_smooth(method = "lm")
Fig8_1
```

However, it is important to realize that the 95% CI reflects our confidence in the 'average' relationship—the regression line itself. In this case we want to use this regression relationship to predict the hardness of new wood samples based on their density. We want to accompany our prediction with an interval that conveys our confidence in it, and the 95% CI does not do that. Instead, we need a prediction interval (PI). The 95% CI can be said to convey our inferential uncertainty, since it conveys our confidence in our estimate of the regression relationship. However, there is scatter around

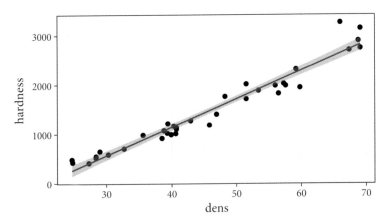

Figure 8.1 Linear regression relationship between timber hardness and wood density with 95% CI.

this line, and because the PI applies to a prediction for a new point (our new wood sample with a known density of 65 but unknown hardness) the 95% PI needs to include this predictive uncertainty as well. The predictive uncertainty reflects the degree of scatter around the line as quantified by the residual variance (the residual or error mean square). For normal least squares linear models, the predict() function will provide the standard error, or the upper and lower bounds of confidence or prediction intervals (95% by default) as follows. First, the point estimate of the predicted hardness of a wood sample with a density value of 65 with its standard error:

```
predict(janka.ls1, newdata = list(dens = 65), se = TRUE)
```

```
## $fit
##          1
## 2577.434
##
## $se.fit
## [1] 53.46068
##
## $df
## [1] 34
##
## $residual.scale
## [1] 183.0595
```

Next, the same point estimate with the upper and lower bounds of a 95% confidence interval (which, as we know, will be approximately ±2 standard errors with $n = 36$):

```
predict(janka.ls1, newdata = list(dens = 65), interval = "confidence")
```

```
##          fit      lwr      upr
## 1 2577.434 2468.789 2686.079
```

Finally, we can obtain a prediction of the timber hardness for a wood sample with a density value of 65 together with a 95% prediction interval as follows:

```
predict(janka.ls1, newdata = list(dens = 65), interval = "predict")
```

```
##          fit       lwr        upr
## 1 2577.434 2189.873 2964.996
```

We can extend this process to generate predictions across the range of the x-variable, which we can then join up to form the upper and lower bounds of a confidence interval region (like the grey confidence band in Fig. 8.1). As we have seen, by default, the predict() function produces a predicted value of the response (hardness) for each value of the explanatory variable (density). However, we have only 36 density values and they are not evenly spread. To be sure to get smooth lines, we need to generate a longer, regular sequence of x-values—say, 100 equally spaced values across the gradient of density:

```
xseq <- seq(from = min(janka$dens), to = max(janka$dens),
    length.out = 100)
```

We can now use the regular sequence in the predict() function as 'new data' in place of the actual values of the explanatory variable:

```
prediction_interval <-
predict(janka.ls1, newdata = list(dens = xseq), interval = "predict")
head(prediction_interval)
```

```
##          fit          lwr        upr
## 1 259.9152 -129.610798 649.4411
## 2 285.7060 -103.305865 674.7179
## 3 311.4969  -77.011353 700.0052
## 4 337.2878  -50.727303 725.3029
## 5 363.0787  -24.453754 750.6111
## 6 388.8695    1.809255 775.9298
```

The predict() function has now produced a matrix of predicted hardness values for each of the regularly spaced density values, together with the upper and lower interval bounds—for a 95% prediction interval here (you can alter it for a confidence interval or to get the value of the standard error as shown above). With the ggplot2 package, all of the variables used in drawing a graph must be supplied as part of a dataframe. We can combine the matrix of predictions with the Janka data to make a new 'fig_data' dataframe:

```
fig_data <- data.frame(xseq, prediction_interval)
head(fig_data)
```

```
##        xseq        fit          lwr       upr
## 1 24.70000  259.9152  -129.610798  649.4411
## 2 25.14848  285.7060  -103.305865  674.7179
## 3 25.59697  311.4969   -77.011353  700.0052
## 4 26.04545  337.2878   -50.727303  725.3029
## 5 26.49394  363.0787   -24.453754  750.6111
## 6 26.94242  388.8695     1.809255  775.9298
```

We now have a long, regularly spaced series of values that we can plot (versus xseq) to draw the regression line and the lower and upper bounds of an interval (fit, lwr, and upr, respectively). However, to substitute them for the observed values of density and hardness we have to give them the same names. We can do this with the rename() function from the dplyr package as follows (with a base-R option shown underneath):

```
library(dplyr)
fig_data <- rename(fig_data, dens = xseq, hardness = fit)
# names(fig_data)[names(fig_data) == 'xseq'] <- 'dens'
head(fig_data)
```

```
##        dens  hardness          lwr       upr
## 1 24.70000  259.9152  -129.610798  649.4411
## 2 25.14848  285.7060  -103.305865  674.7179
## 3 25.59697  311.4969   -77.011353  700.0052
## 4 26.04545  337.2878   -50.727303  725.3029
## 5 26.49394  363.0787   -24.453754  750.6111
## 6 26.94242  388.8695     1.809255  775.9298
```

To finish, let's replot the Janka data and compare the 95% confidence and prediction internals. The following code redraws the scatter plot with

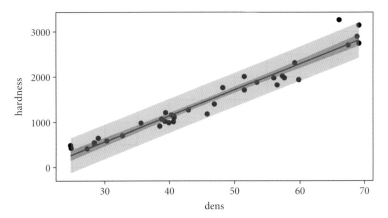

Figure 8.2 Linear regression relationship between timber hardness and wood density with 95% confidence (blue) and prediction (grey) intervals.

the confidence interval (in blue) before using the 'fig_data' dataframe of predicted values to plot the prediction interval (Fig. 8.2):

```
Fig8_2 <- qplot(data = janka, x = dens, y = hardness) +
  geom_smooth(data = janka, method = "lm", fill = "blue") +
  geom_smooth(data = fig_data, aes(ymin = lwr, ymax = upr),
    stat = "identity")
Fig8_2
```

The take-home message is how much wider the prediction interval is than the confidence interval. As the old quip goes, prediction is hard... especially about the future.

8.4 Summary: Statistics

Once the values of the intercept and slope have been estimated, linear regression models can be used to predict values of *y* from *x*—in this case to estimate timber hardness based on its relationship with wood density. The uncertainty in the predicted values can be quantified by a prediction interval. These are similar to confidence intervals, but they are wider because they contain the inferential uncertainty about the regression line that is captured by the confidence interval plus the predictive uncertainty due to the scatter of points around the regression line.

8.5 Summary: R

- Prediction from linear models is done using the predict() function.
- The predict() function provides point estimates together with the standard error or a confidence or prediction interval.
- A nice feature of R is the ability to see the model matrix via the function of the same name.

Testing

9.1 Significance testing: Time for *t*

Over recent years statisticians have pointed out that most scientists are overly fixated on *P*-values and that this has contributed to the reproducibility crisis that we find ourselves in. As a response, we have so far preferred an estimation-based approach. Working with estimates and confidence intervals has the advantage of keeping us closer to the science that we are interested in—no scientist studies *P*-values in their own right! However, sometimes you may really feel like you need a *P*, if only because a journal or colleague is demanding it. Luckily, there are significance tests we can apply to the coefficients of a linear model. This chapter introduces one of the classics—Student's *t*-test.

9.1.1 R PACKAGES

```
library(arm)
library(ggplot2)
library(Sleuth3)
library(SMPracticals)
```

The New Statistics with R: An Introduction for Biologists. Second Edition. Andy Hector,
Oxford University Press. © Andy Hector 2021. DOI: 10.1093/oso/9780198798170.003.0009

9.2 Student's *t*-test: Darwin's maize

Student's *t*-test was invented by William Sealy Gosset. The test uses the *t*-distribution, which can be thought of as a small-sample-size version of the normal distribution. Gosset needed this small-sample-size alternative because he was performing many experimental comparisons, sometimes with very small sample sizes (Box 9.1).

Box 9.1 - Student's *t*-test and Guinness beer

Student's *t*-test got its name because its inventor, William Sealy Gosset (1876–1937), published it using the pseudonym 'Student' (Senn 2008). Gosset was Head Brewer for Guinness (maker of the eponymous stout) and conducted research on how to improve the brewing of their beers, including trials of different combinations of yeast, hops, and barley. Sample sizes were sometimes as small as four replicates for a particular combination. Gosset wrote under a false name because Guinness employees were not allowed to publish under their own. Despite inventing the test, Gosset apparently did not place too much weight on statistical significance. For him (and Guinness), what mattered more was the practical significance of the results in terms of the beer that was produced—how much it cost, how it tasted, and so on (Ziliak & McCloskey 2008). One of the key aims of this book is to pass that point on: biological significance (via the estimates and intervals and what can be done with them) matters more than statistical significance and should be our primary focus.

The *t*-test comes in two basic types depending on the design of the study and the resulting form of the data being analysed. The one-sample *t*-test takes the mean of a single sample and compares it with a null hypothesis of zero, while the two-sample *t*-test compares the difference between the means of two samples against a null hypothesis of no (zero) difference. As the name implies, the paired two-sample *t*-test is a subspecies that applies when the values of two samples come in pairs (as in the Darwin's maize example). R has a function designed to apply the *t*-test, but I often avoid it when teaching and am not going to use it here (Box 9.2).

Box 9.2 - R's *t*-test function

As we'll see below, the ingredients of the *t*-test are a difference, its standard error (and the confidence intervals calculated using it), the observed value of *t*, the critical value of *t*, the degrees of freedom, and the *P*-value. From a teaching perspective, the R *t*-test function output is poorly ordered and missing some of these bits of information, in particular the standard error (to its credit, it does give a 95% confidence interval that you could back-calculate the SE from). The default setting is also not the classic *t*-test but a later extension by Welch—this may be good for research (Welch's *t*-test does not assume equal variances) but it adds a further twist to explain to beginners. It also presents only the difference and nothing about the quantities used to calculate it (the mean crossed and selfed plant heights in this case). Finally, good luck deciphering the help information!

The general form of the *t*-test is

$$t = \frac{\text{difference}}{\text{SE}}$$

So, the calculation of the *t*-value is a more formal and exact version of what we were doing earlier in the book when we inspected the output of the display() function and compared the estimates with their standard errors. We are essentially counting in standard errors: our rough rule of thumb has been that estimates about twice as large as their standard error are notable at our lowest level of confidence (95%), estimates about three times as large are notable at the next level (99%), and so on. Of course, these are approximate 'eyeball' tests, and they assume that we can use a normal distribution, an approximation that becomes poorer as samples become smaller. When sample sizes are larger, *t* converges to the normal distribution, but when they are small the *t*-distribution is shorter and wider than normal. Figure 9.1 compares *t*-distributions for Darwin's maize data with $n = 15$ pairs and Gosset's smallest trials with $n = 4$.

One potentially confusing aspect of the *t*-test is that it involves two *t*-values. One is the *critical* value of *t*, which 'sets the bar' for the comparison—this is the minimum *t*-value needed to achieve a given level of significance ($P = 0.05$, 0.01, etc.). The second is the *observed* value of *t*, calculated by dividing the estimate by its standard error. When the observed value of *t* is larger than the critical value of *t*, the result is declared statistically significant at that level. Of course, declaring values significant at a given level is a hangover from the early days when computing exact

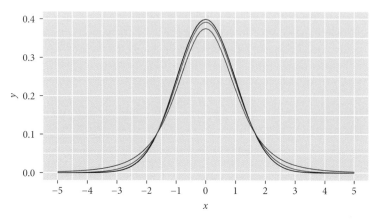

Figure 9.1 The standard normal distribution (black) compared with the *t*-distribution for samples of size *n* = 14 (red) and *n* = 4 (blue).

P-values was demanding and analysts used tables of critical values as a shortcut. Nowadays, statistical software gives us the exact values (which we should report).

Although it is a classic test and often one of the first taught to beginners, in some ways the *t*-test brings a lot of extra complexity to the early stages of introductory statistics courses. I am also not sure how often *t* is really essential nowadays in this age of 'big data'. Gosset needed *t* as his sample sizes were sometimes very small (*n* = 4!), but it is not clear how often modern analysts are in the same position, especially with statisticians advising we avoid very small sample sizes. For example, so far we have followed Gelman and Hill and colleagues in taking a (refreshingly) casual approach that is often happy (for introductory teaching purposes at least) to calculate approximate 95% CIs as ±2 SEs, but if we use the *t* distribution instead of assuming we can apply the normal approximation, what difference does it make? Figure 9.2 plots the critical value of *t* (for *P* = 0.05) as a function of sample size, showing that it is only when sample sizes get down into single digits that *t* starts to get much larger than our rough large-sample approximation (the blue dashed line). The values are as follows (printed across the page to save space):

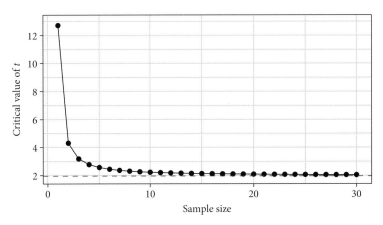

Figure 9.2 Critical value of *t* as a function of sample size.

```
##   [1] 12.706205  4.302653  3.182446  2.776445  2.570582
##   [6]  2.446912  2.364624  2.306004  2.262157  2.228139
##  [11]  2.200985  2.178813  2.160369  2.144787  2.131450
##  [16]  2.119905  2.109816  2.100922  2.093024  2.085963
##  [21]  2.079614  2.073873  2.068658  2.063899  2.059539
##  [26]  2.055529  2.051831  2.048407  2.045230  2.042272
```

Darwin's maize data has a sample size of 30, generated by 15 pairs of crossed and selfed plants. In Chapter 5 we have already calculated the 15 differences in height, calculated the mean difference in height and its standard error, and formed an approximate 95% CI as ±2 SEs. To make this a paired *t*-test we simply compare the observed value of *t* (the mean divided by the SE) with the critical value of *t* (rather than approximating it by a value of 2). We can see from Fig. 9.2 that the critical value is a little larger than 2. The qt() function provides quantiles that can be used to obtain the exact *P*-value; in this case (for $P = 0.05$ and $n = 15$ pairs) the critical value of *t* is

```
qt(0.975, df = 14)
```

```
## [1] 2.144787
```

So far, we have mainly used the arm package display() function to look at the coefficients. The arm display() function was written for teaching

purposes as a simpler and more compact version of the base-R summary()
function. If we switch to the summary() function, we can see that it includes
t-tests for the coefficients. Let's revisit Darwin's maize as an example (to
keep things simple, let's start with the simplest linear model, which omits
the pairing):

```
summary(lm(height ~ type, data = darwin))
```

```
##
## Call:
## lm(formula = height ~ type, data = darwin)
##
## Residuals:
##      Min      1Q  Median      3Q     Max
## -8.1917 -1.0729  0.8042  1.9021  3.3083
##
## Coefficients:
##               Estimate Std. Error t value Pr(>|t|)
## (Intercept)    20.1917     0.7592  26.596   <2e-16 ***
## typeSelf       -2.6167     1.0737  -2.437   0.0214 *
## ---
## Signif. codes:
## 0 '***' 0.001 '**' 0.01 '*' 0.05 '.' 0.1 ' ' 1
##
## Residual standard error: 2.94 on 28 degrees of freedom
## Multiple R-squared:  0.175,  Adjusted R-squared:  0.1455
## F-statistic:  5.94 on 1 and 28 DF,  p-value: 0.02141
```

To save us having to specify the tests we want to perform (although
perhaps that would be good for us!), a test is automatically performed for
every row of the table. Sometimes these are the tests planned *a priori* that
we are interested in, and sometimes not. What null hypothesis does the first
row of the table test? And, is it a test we had planned to make in advance?
For example, the test in the second row is an *a priori* comparison—it
tests our null hypothesis by comparing the observed difference in height
between the selfed and cross-pollinated plants (−2.6 inches) with a null

hypothesis of zero difference. However, the first row tests the mean height of the cross-pollinated plants versus a null hypothesis of an average height of zero. This was not a comparison we intended to make in advance, nor is it of interest *post hoc* to know if plant height is significantly different from zero. Although it would mean more typing, in some ways it would be good for us to have to specify only the tests we wanted to perform rather than every row being automatically tested (indeed, that is how the lme4 package currently works—but to the annoyance of many users!). The average difference in height is −2.6167 ± 1.0737, producing an observed value of *t* of

```
-2.6167/1.0737
```

```
## [1]  -2.437087
```

Of course, as usual, it is arguably better to work with the more informative confidence intervals:

```
confint(lm(height ~ type, data = darwin))
```

```
##                    2.5 %      97.5 %
## (Intercept)  18.63651  21.7468231
## typeSelf      -4.81599  -0.4173433
```

While it is a relatively easy to work with this table of coefficients, this is the table for the standard two-sample *t*-test, not for the paired version. To get the equivalent of a paired *t*-test, we simply have to add the factor for the pairs to the linear-model formula:

```
summary(lm(height ~ type + pair, data = darwin))
```

```
##
## Call:
## lm(formula = height ~ type + pair, data = darwin)
```

```
##
## Residuals:
##     Min      1Q  Median      3Q     Max
## -5.4958 -0.9021  0.0000  0.9021  5.4958
##
## Coefficients:
##                Estimate Std. Error t value Pr(>|t|)
## (Intercept)    21.7458     2.4364   8.925 3.75e-07 ***
## typeSelf       -2.6167     1.2182  -2.148   0.0497 *
## pair2          -4.2500     3.3362  -1.274   0.2234
## pair3           0.0625     3.3362   0.019   0.9853
## pair4           0.5625     3.3362   0.169   0.8685
## pair5          -1.6875     3.3362  -0.506   0.6209
## pair6          -0.3750     3.3362  -0.112   0.9121
## pair7          -0.0625     3.3362  -0.019   0.9853
## pair8          -2.6250     3.3362  -0.787   0.4445
## pair9          -3.0625     3.3362  -0.918   0.3742
## pair10         -0.6250     3.3362  -0.187   0.8541
## pair11         -0.6875     3.3362  -0.206   0.8397
## pair12         -0.9375     3.3362  -0.281   0.7828
## pair13         -3.0000     3.3362  -0.899   0.3837
## pair14         -1.1875     3.3362  -0.356   0.7272
## pair15         -5.4375     3.3362  -1.630   0.1254
## ---
## Signif. codes:
## 0 '***' 0.001 '**' 0.01 '*' 0.05 '.' 0.1 ' ' 1
##
## Residual standard error: 3.336 on 14 degrees of freedom
## Multiple R-squared:  0.469,  Adjusted R-squared:  -0.09997
## F-statistic: 0.8243 on 15 and 14 DF,  p-value: 0.6434
```

The table of coefficients gets more complex because it takes the crossed plant from pair 1 as the intercept, shows the mean difference in height of the crossed plants (−2.6167), and then shows the average difference of each pair relative to pair 1. But, if we ignore the differences among pairs, the second row now gives us the paired t-test:

```
-2.6167/1.2182
```

```
## [1] -2.148005
```

For completeness, we can use the linear-model function to perform the equivalent of a one-sample *t*-test. First we need a single sample of differences—let's use the differences in plant height for Darwin's maize from Chapter 5:

```
library(Sleuth3)
ex0428$Difference <- ex0428$Cross - ex0428$Self
```

To fit a linear model that just estimates the mean difference, we use a model formula with just a '1' to indicate an intercept—which will be the mean of the single sample of differences (this is the same type of intercept-only model as 'ls0' from Chapter 6). The summary() function output contains a one-sample *t*-test of the mean difference in height:

```
summary(lm(Difference ~ 1, data = ex0428))
```

```
##
## Call:
## lm(formula = Difference ~ 1, data = ex0428)
##
## Residuals:
##      Min       1Q   Median       3Q      Max
## -10.9967  -1.2417   0.3833   3.0083   6.7633
##
## Coefficients:
##              Estimate Std. Error t value Pr(>|t|)
## (Intercept)     2.617      1.219   2.147   0.0498 *
## ---
## Signif. codes:
## 0 '***' 0.001 '**' 0.01 '*' 0.05 '.' 0.1 ' ' 1
##
## Residual standard error: 4.719 on 14 degrees of freedom
```

9.3 Summary: Statistics

The *t*-test and the *t*-distribution it is based on come into their own when sample sizes are very small. Although the *t*-test is one of the classics, it was probably more often needed in the early twentieth century than it is in this age of larger (sometimes huge) data sets. Nevertheless, we may sometimes find ourselves in need of *t*. R, has dedicated functions for doing *t*-tests but, as shown above, it is easy to fit linear models that produce one-sample, two-sample, and paired versions of the *t*-test in the summary() table output. As the sample size increases, the *t*-distribution quickly converges towards the normal distribution and we can keep things simple and use the normal-distribution approximation (±2 SEs for an approximate 95% CI, etc.).

9.4 Summary: R

- The summary() function contains *t*-tests for the coefficients of a linear model.
- The qt() function provides quantiles that can be used to obtain exact *P*-values.

9.5 References

Senn, S. (2008) A century of *t*-tests. *Significance* 5: 37–9.
Ziliak, S.T. & McCloskey, D.N. (2008) *The Cult of Statistical Significance*. University of Michigan Press.

Intervals

10.1 Comparisons using estimates and intervals

The reproducibility crisis has many causes, but statisticians tell us one contributory factor is our overuse of null hypothesis significance testing and our unhealthy fixation on probability values. Ironically, this leads us to pay too little attention to the estimates and intervals that summarize the biological effects (and our degree of confidence in them) that should be of greater interest. One simple action we can take is to prefer an estimation-based approach to analysis that focuses on working as much as possible with the estimates and intervals and demotes the use of P-values to a more appropriate level (Cumming 2013). This chapter gives a brief introduction and user's guide to intervals and how to apply them in linear-model analysis. We have already met most of the summary statistics and intervals in previous chapters, but I am pulling them together here to allow us to compare and contrast the different approaches and to re-emphasize them over significance testing.

The New Statistics with R: An Introduction for Biologists. Second Edition. Andy Hector,
Oxford University Press. © Andy Hector 2021. DOI: 10.1093/oso/9780198798170.003.0010

10.1.1 R PACKAGES

```
library(arm)
library(cowplot)
library(ggplot2)
library(Sleuth3)
library(SMPracticals)
```

10.2 Estimation-based analysis

With the significance-testing approach covered in earlier chapters, we base our conclusions on the signal-to-noise ratio of the ANOVA table F-tests or the summary table t-tests, and whether these pass some arbitrary level of statistical significance. Significance testing can be a powerful approach when properly applied, but statisticians tell us we are overusing it in a scatter-shot way, peppering our papers with P-values. Significance testing has other downsides too—it draws us into a binary way of thinking that categorizes results as significant or not in relation a single arbitrary dividing line. Many statisticians are encouraging us to curb our overuse of P-values and to instead draw our inferences based on point estimates and intervals. Because these estimates and intervals are on the original scale of measurement, they keep us more closely connected to the questions that originally motivated our analysis and present statistical results in a way that makes it easier to judge their scientific importance rather than just their statistical significance. In fact, the use of estimates and intervals has several advantages over the focus on P-values. However, the number of different types of error bars and intervals can be confusing. My experience of teaching statistics and as a journal editor and reviewer leads me to suspect that the understanding of error bars and intervals is generally far from perfect (I am being diplomatic here!). The obsession with adding letters to graphs to show which pairs of means are significantly different (or not) suggests that the reader has no idea what the error bars are telling them (these letters often redundantly repeat what is obvious from the error bars and have all of the significance-testing downsides mentioned above and more). It's amazing how often papers apply a complex statistical modelling approach and yet the presentation of the results suggests a lack of a basic understanding of error bars and confidence intervals.

We can divide the different options into two classes—descriptive and inferential (although not everyone likes this division). As in the rest of this book, I focus on the general principles and try to keep things as simple and straightforward as possible by avoiding unnecessary complexities that could obscure the bigger picture. For example, here I sometimes assume large sample sizes so that we don't have to revisit the complexities of the t-distribution covered in the last chapter (t only gets appreciably larger than 2 once we get down to very small sample sizes). I also assume that sample sizes are equal between treatments (which simplifies the relationship between the standard error of the mean and the standard error of the difference).

10.3 Descriptive statistics

In Chapter 3 we applied a range of descriptive approaches to Darwin's maize data. With small data sets it can be good to simply plot the individual values, while if the group size is large enough boxplots provide a richer summary (the boxplots display five estimates, so you need sample sizes larger than that for each group being summarized). Other modern alternatives include violin plots. To save space, these graphs are not repeated here; please refer back to Chapter 3.

10.3.1 ERROR BARS

The other common type of graph that we have already met presents a measure of central tendency—like the mean—with a measure of variability. Sometimes the mean is indicated by the height of a column, but such bar charts are wasteful (since only the top of the bar carries any information) and can sometimes be visually misleading (Wickham 2016). It is better to indicate the mean with a symbol and indicate the variability using 'error bars'. There is no single 'best' type of error bar—it depends what you want to show to the reader.

10.3.2 STANDARD-DEVIATION ERROR BARS

The classic descriptive statistics for continuous data (like Darwin's maize data) are the mean and standard deviation. We have already presented the means, together with standard deviations that were calculated separately for each group, but there is an alternative we can demonstrate here. Linear-model analysis calculates a 'pooled' estimate of the residual variability across all treatments (we will discuss this in greater detail in the next chapter in relation to ANOVA tables), which can be used instead of calculating separate SDs for each group. The standard deviation is the first of several types of error bar that we are going to compare, so we need a general approach to plotting them. Because there are so many different types of interval we normally need to use R to do the calculations ourselves and then provide the resulting values to the graphics function to draw the error bar. There are several alternative approaches even within the ggplot2 package, but here we are going to use the flexible geom_linerange() function. Because we calculate the values for the ends of the lines drawn by linerange(), this function can be used to draw any type of error bar or interval. First, we need to re-fit the linear model of Darwin's data that will provide the estimate of the pooled standard deviation:

```
ls1 <- lm(height ~ type, data = darwin)
```

We are going to draw a series of graphs that present the pollination treatment means together with error bars that show a variety of different calculated intervals. Remember that the ggplot2 package needs all of the ingredients for a graph to be in a dataframe, so let's set one up ready to hold the calculated values that we want to plot (this is an organized way to work even when not using ggplot2). One way to do do this is to use the expand.grid() function to make a dataframe based on the levels of the explanatory variable (pollination type):

```
estimates <- expand.grid(type = levels(darwin$type))
```

Before we look at the 'estimates' object, let's add a column (height) for the treatment means (which we will plot as point symbols in due course). We can generate the mean values using either the predict() function (supplying our two-row dataframe 'estimates' using the newdata argument) or the table-apply function, both of which we have already met and used:

```
estimates$height <- predict(ls1, newdata = estimates)
# estimates$height <- tapply(darwin$height,
# darwin$type, mean)
estimates
```

```
##     type   height
## 1 Cross 20.19167
## 2  Self 17.57500
```

Earlier, we saw that the output of the (arm package) display() function contains the estimate of the pooled standard deviation that we want to display as an error bar. We can create ('lmDisplay') an object from the display() function output and extract the pooled SD, which it calls 'sigma.hat':

```
lmDisplay <- display(lm(height ~ type, data = darwin))
```

```
## lm(formula = height ~ type, data = darwin)
##             coef.est coef.se
## (Intercept) 20.19     0.76
## typeSelf    -2.62     1.07
## ---
## n = 30, k = 2
## residual sd = 2.94, R-Squared = 0.18
```

```
lmDisplay
```

```
## $call
## lm(formula = height ~ type, data = darwin)
##
## $sigma.hat
## [1] 2.94038
##
## $r.squared
## [1] 0.175003
##
## $coef
## (Intercept)     typeSelf
##   20.191667    -2.616667
##
## $se
## (Intercept)     typeSelf
##   0.7592028    1.0736749
##
## $t.value
## (Intercept)     typeSelf
##   26.595880    -2.437113
##
## $p.value
##  (Intercept)      typeSelf
## 2.049810e-21 2.141448e-02
##
## $n
## [1] 30
##
## $k
## [1] 2
```

We can now add this to our table (repeating the value in each of the two rows):

```
estimates$SD <- rep(lmDisplay$sigma.hat, 2)
estimates
```

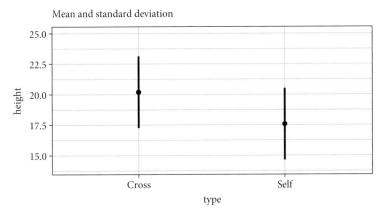

Figure 10.1 Treatment means with the pooled standard deviation.

```
##    type    height        SD
## 1 Cross 20.19167 2.94038
## 2  Self 17.57500 2.94038
```

Now we have all the necessary ingredients for the graph (Fig. 10.1):

```
Fig10_1 <-
  qplot(data = estimates, x = type, y = height) +
  ylim(14, 25) +
  ggtitle("Mean and standard deviation") +
  geom_linerange(aes(ymin = height - SD, ymax = height +
    SD), size =1)
Fig10_1
```

Descriptive statistics do just that—they give a descriptive summary of the central tendency and variability. But, to address scientific questions, we often need *inferential statistics* that allow us to infer how different experimental treatments are relative to the background noise.

10.4 Inferential statistics

10.4.1 STANDARD ERROR BARS

Recall from earlier that we can think of standard errors as the standard deviations of estimated statistics—in this case the SEM is the standard deviation of the estimated means. The interval reflects uncertainty due to

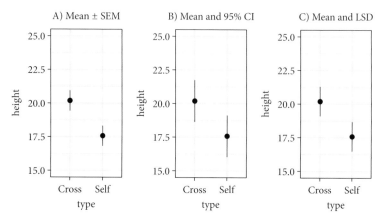

Figure 10.2 Treatment means with (a) SEMs, (b) 95% confidence intervals, (c) LSDs.

sampling (different samples would give different estimates) and conveys our confidence in where the 'true' mean probably lies. The use of plots of means and SEMs is a common approach because the standard error is the basis for many other inferential statistics—the other error bars we will go on to calculate are all multiples of the SE. First, we need to use the predict() function to extract the value of the standard error (I won't keep reprinting the updated 'estimates' table, to save space, but you may want to take a look at it at each step):

```
estimates$SEM <- predict(ls1, newdata = estimates,
    se.fit = TRUE)$se.fit
```

Then we can draw the graph of the means with the SEMs (Fig. 10.2(a)):

```
Fig10_2a <-
  qplot(data = estimates, x = type, y = height) +
  ylim(15, 25) +
  ggtitle("A) Mean ± SEM") +
  geom_linerange(aes(ymin = height - SEM, ymax = height + SEM),
    size = 1)
Fig10_2a
```

10.4.2 CONFIDENCE INTERVALS

The closest thing to a general-purpose interval is the 95% confidence interval—this is what statisticians are encouraging us to present in most cases. These graphs show the point estimates ±2 standard errors (when sample sizes are small, the *t*-distribution will increase the number of standard errors we need to achieve the same level of confidence, as we saw in the last chapter). Confidence intervals both express our confidence in the estimate and allow hypothesis tests: values falling inside the interval are consistent with the data, and those falling outside it are not (a significant difference equivalent to a $P < 0.05$ result produced by a *t*- or *F*-test, as we have seen in previous chapters). So, confidence intervals can be used to test whether observed means are significantly different from hypothesized values (often zero, but any value in principle). In this way, the confidence intervals do everything the *t*- or *F*-test does plus more (although that means they also suffer from some of the same drawbacks). Because the interval tells us the whole range of mean values consistent with the data, the reader is also able to test whether any other value of interest lies inside the interval or not.

We could just tweak the last chunk of R code to plot approximate 95% CIs as ±2 SEMs, but the predict() function will supply exact values (calculated with the appropriate critical value of *t*) of the upper and lower interval bounds (note the use of the square indexing brackets to extract the second and third values and that we specify a confidence interval, not a prediction interval):

```
estimates$CI95_lwr <- predict(ls1, newdata = estimates,
    interval = "confidence")[, 2]
estimates$CI95_upr <- predict(ls1, newdata = estimates,
    interval = "confidence")[, 3]
```

Then we just substitute the new values into our generic qplot() code for the point-and-interval plot (Fig. 10.2(b)):

```
Fig10_2b <-
    qplot(data = estimates, x = type, y = height) +
    ylim(15, 25) +
    ggtitle("B) Mean and 95% CI") +
    geom_linerange(aes(ymin = CI95_lwr, ymax = CI95_upr), size = 1)
Fig10_2b
```

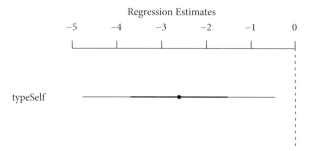

Figure 10.3 Difference in mean height with approximate 95% CI.

One limitation of graphs showing standard errors or 95% confidence interval bars is that they do not show exactly when two means are statistically significant or not (although a savvy reader can usually judge approximately). If we want a graph that unambiguously displays when means are significantly different or not, there are two main options. The first is to plot the difference between means with a confidence interval for this mean difference. The second is to plot the two means with error bars that show the wonderfully named least significant difference (LSD).

10.4.3 CONFIDENCE INTERVALS FOR DIFFERENCES BETWEEN MEANS

One way to see whether a difference between means is significant or not is to plot the difference with a confidence interval calculated using the standard error of the difference. Working with the mean difference and the SED is best when dealing with matched data (like paired data). The arm package has a coefplot() function for generating this type of graph (Fig. 10.3):

```
library(arm)
fig10_3 <- coefplot(ls1, xlim = c(-5, 0))
```

The graph shows the difference in height (-2.6 inches) $\pm$ 1 and 2 SEMs (the approximate 68% and 95% CIs for the difference in means). As we saw in Chapter 6, because zero difference is outside the confidence interval, it is not consistent with the data and we can reject it at this level of confidence (the xlim argument is used only to extend the *x*-axis limits to show zero). The disadvantage of this type of graph is that we see only the difference in the means, not the means themselves. The LSD solves this problem by showing the means while using the SED to construct intervals that show whether they are significantly different or not.

10.4.4 LEAST SIGNIFICANT DIFFERENCES

The LSD is equivalent to a 95% CI for the difference between means since the LSD equals $t \times$ SED (approximately $2 \times$ SED with a large sample size). The LSD can be a bit confusing because, rather than showing the difference between means (as coefplot does), it shows the two means. Instead of comparing one estimate with a null value of zero using a single interval, it shows both means and we have to 'share' the interval between them. To do that, we centre the LSD interval on each mean so that an overlap indicates means that are not significantly different. First, we need to extract the SED from the linear model:

```
estimates$SED <- summary(ls1)$coefficients[2, 2]
estimates
```

```
##     type   height      SD         SEM CI95_lwr CI95_upr
## 1 Cross 20.19167 2.94038 0.7592028 18.63651 21.74682
## 2  Self 17.57500 2.94038 0.7592028 16.01984 19.13016
##         SED
## 1 1.073675
## 2 1.073675
```

With this unpaired version of the analysis we have 28 degrees of freedom, which means the value of *t* is very close to 2:

```
qt(0.975, df = 28)
```

```
## [1] 2.048407
```

Importantly, because we are comparing two means, each with an interval, we centre the LSD interval on each mean—in other words, we show each estimate ± half of the least significant difference:

```
Fig10_2c <-
  qplot(data = estimates, x = type, y = height) +
  ylim(15, 25) +
  ggtitle("C) Mean and LSD") +
  geom_linerange(aes(ymin = height - 1.02 * SED, ymax = height +
    1.02 * SED), size = 1)
Fig10_2c
```

The plot_grid() function from the cowplot package provides one way to put the plots side by side (as in Fig. 10.2) for comparison:

```
library(cowplot)
Fig10_2 <-
  plot_grid(Fig10_2a, Fig10_2b, Fig10_2c, ncol = 3,  nrow = 1)
Fig10_2
```

10.4.5 MULTI-INTERVAL PLOTS

One advantage of an estimation-based approach using confidence intervals is that it provides a more continuous view (as opposed to the dichotomous impression given by significance tests). We can enhance this by plotting multiple intervals with different levels of significance (or, similarly, with bars showing different multiples of the standard error) (Fig. 10.4(a)):

```
Fig10_4a <-
  qplot(data = estimates, x = type, y = height) +
  ylim(11, 27) +
  geom_point(shape = 3) +
  ggtitle("A) Multi-interval plot") +
  geom_linerange(aes(min = height - 1 * SEM,
    max = height + 1 * SEM), size = 1.4) +
  geom_linerange(aes(min = height - 2 * SEM,
    max = height + 2 * SEM), size = 0.7) +
  geom_linerange(aes(min = height - 3 * SEM,
    max = height + 3 * SEM), size = 0.3)
Fig10_4a
```

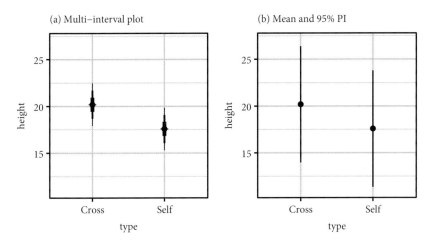

Figure 10.4 (a) Mean heights ± 1, 2, and 3 SEMs; (b) mean heights with a 95% prediction interval.

10.4.6 PREDICTION INTERVALS

In Chapter 7 we explored the distinction between a confidence interval and a prediction interval. To extract the upper and lower bounds for a 95% prediction interval we simply change the interval argument from "confidence" to "prediction", saving the values (as P95_lwr and P95_upr) and substituting these into the qplot() code from above. Remember that the prediction interval is wider than the confidence interval, substantially so in this case (Fig. 10.4(b)):

```
Fig10_4 <- plot_grid(Fig10_3a, Fig10_4b, ncol = 2, nrow = 1)
Fig10_4
```

10.5 Relating different types of interval and error bar

The SEM is the basis for intervals relating to means, and the SED plays the same role when the focus is on differences between means. Therefore, if we know how the SED is related to the SEM, we know how different intervals are related to each other too. For example, we can mentally sketch approximate LSDs onto graphs showing means and SEMs (or means and CIs). Figure 10.4(a) plots 95% CIs as the wider intervals (drawn with thin lines) and LSDs as the narrower intervals. The point is to imagine that we have only been presented with the 95% CIs but, because we know how the SEM and SED are related, we can estimate approximate LSDs ourselves. The three panels of Fig. 10.5 show scenarios where we know the means are clearly not significantly different at $P = 0.05$ (Fig. 10.5(a)), where the means are significantly different at approximately $P = 0.01$ (Fig. 10.5(b)), and where they are different at about $P = 0.05$ (Fig. 10.5(c)).

The rules of thumb presented in Fig. 10.5 work because (at the $P < 0.05$ level of significance and with large sample sizes) the SED is approximately 1.4 SEMs, making the LSD intervals approximately 2.8 SEMs wide. Because the LSD interval is centred on each mean, the upper and lower halves of the LSD bars are approximately 1.4 SEMs long. In other words, for a rough eyeball test the bounds of an LSD interval lie just under midway between the ends of a conventional error bar (which shows ±1 SEM) and the ends of a 95% CI (which shows ±2 SEM). Knowing how one interval relates to another therefore enables us to perform rough eyeball tests using the following rules of thumb. For example, when comparing two means with 95% CIs, if one mean is contained within the interval around the other mean then the means are not significantly different at

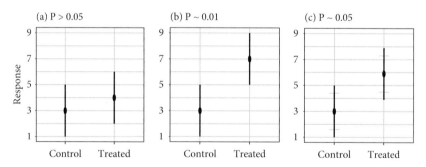

Figure 10.5 Rules of thumb for inferring approximate statistically significant means from graphs presenting 95% CIs. (a) Non-significant ($P > 0.05$); (b) significant ($P \sim 0.01$); (c) significant ($P \sim 0.05$; grey crossbars indicate the limits of least significant difference intervals).

$P = 0.05$ (Fig. 10.5(a)). If the intervals do not overlap at all then the means are significantly different (95% CIs that are at the point of overlap are roughly equivalent to means that are significantly different at $P = 0.01$; Fig. 10.5(b). In situations between these two extremes it is still possible to judge approximate statistical significance by knowing that LSD intervals plot the mean plus and minus approximately 1.4 SEM. For example, in Fig. 10.5(c) it is possible to judge by eye that the 95% confidence intervals overlap by roughly half their length, which equates to approximate significance at $P < 0.05$ (the grey crossbars mark the limits of the LSD intervals that visually indicate this).

10.5.1 INTERPRETING CONFIDENCE INTERVALS

Appreciating the advantages of the estimation-based approach over null hypothesis significance testing is not as easy as it sounds. The following hypothetical scenarios may help to bring out some of the subtleties. Imagine a situation where we are analysing a response variable and where we have a hypothesized null value of zero. Surely, given the set-up, this is a case where we should follow a null-hypothesis-testing approach? No; even in this case there are potential advantages to focusing on the estimates and intervals. Consider three different outcomes (Fig. 10.6).

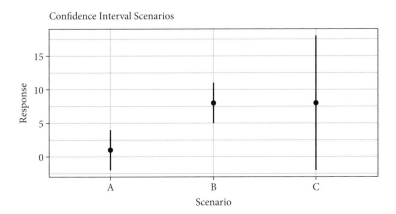

Figure 10.6 Three different confidence interval scenarios.

In scenario A, the mean is close to zero, which lies well inside the 95% CI, and a significance test (a one-sample t-test for example) versus a null hypothesis of zero would produce a clearly non-significant result. In scenario B, the width of the CI is the same but the mean value is much larger, so that zero is clearly outside the interval and a significance test would produce a clear positive result. Finally, in scenario C, the mean value is the same as in B but now the interval is much wider (it has the same lower bound as in A) and includes zero; a test would give a non-significant result. First, imagine using a null hypothesis significance-testing approach (for example, based on the output of three one-sample t-tests). This approach steers us towards dichotomous thinking in which scenarios A and C have the same outcome and only B has an effect. However, if we instead plot estimates and intervals we are less inclined to such thinking.

The outcomes of scenarios A and B are pretty clear-cut, and significance testing and estimation-based analyses would lead us to the same conclusions. However, the subtleties of scenario C are more apparent when considering estimates and intervals. It is true that the interval for C includes zero, but the estimated mean value is as large as in scenario B; there is just much greater uncertainty. However, this means that even larger values (of up to 18) are plausible. So, while this result is non-significant by the convention of $P < 0.05$, there is the possibility of a large effect (as large as or larger than that in scenario B). The more continuous view conveyed by the confidence intervals is much more likely to lead us to think this way and look at the sample size for scenario C, and consider repeating the study with a larger sample size if necessary.

10.5.2 POINT ESTIMATES AND CONFIDENCE INTERVALS FOR RESEARCH SYNTHESIS AND META-ANALYSIS

Scenario C is also the type of 'boring', 'unclear' non-significant result that (given the preferences of most journals, editors, and reviewers) has a lower chance of being published (or of being clearly reported even if it is published) and that is therefore less likely to be included in subsequent meta-analysis. However, it is important that any meta-analysis should include all of the research conducted on a topic (given sufficient quality etc.), both significant and non-significant. Imagine there are several cases like scenario C in the literature where the intervals are wide, in part because those studies had small sample sizes. By combining data from all studies, meta-analysis can detect general effects that may have been declared to be non-significant (or marginal) in each of the smaller, original studies. Repeatability of a result is a key part of the scientific method (one which, sadly, our emphasis on novelty works against), and meta-analysis is one of our main tools for detecting repeatable results. However, data can only be included in a meta-analysis if they are available, and presenting point estimates, intervals, and sample sizes for all results (whether statistically significant or not) helps enormously in facilitating these syntheses of published research. This is yet another reason for favouring an estimation-based approach to analysis.

As just mentioned, the repeatability of results is a key part of the scientific method. Unfortunately, the premium placed on novelty by both journals and the grant review process works against the demonstration of repeatability. Nevertheless, generally similar experiments are eventually performed, and as results build up in the literature they can be synthesized to see how consistent the results are and how contingent the outcome is on confounding variables. There are ways of working with P-values for both formal meta-analysis and less formal reviews of research, but effect sizes with a clearly defined interval (and sample sizes) are vastly superior and preferable for both. Less formal reviews of research often use 'vote counting', where each analysis contributes a statistically significant or non-significant result. For example, our analysis of Darwin's data produced a statistically significant result (just) and would count as a positive vote for negative effects of inbreeding depression on fitness. Now consider another (imaginary) study—say by Wallace—with a similar mean effect of inbreeding but which was not statistically significant. Focusing on P-values gives us a picture of inconsistency, since Darwin's study was statistically significant but Wallace's was not—one vote each way. Once again, it is preferable to present point estimates and intervals. Figure 10.7 displays both results, and gives a different impression.

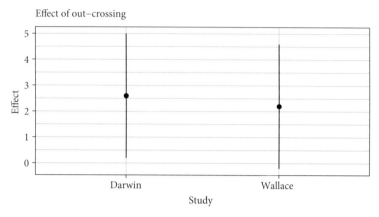

Figure 10.7 Hypothetical confidence interval scenarios.

We can see in Fig. 10.7 that one result is statistically significant (the interval excludes zero) and one is not, but it is also much clearer that the outcomes are actually very similar, with one interval just containing zero and the other just excluding it. But, again, let us not focus too narrowly on statistical significance. More importantly, both studies produced positive effects—in fact positive effects of a consistent size and with very similar degrees of uncertainty (assuming Wallace used similar sample sizes). Once again, the estimates and intervals are superior for conveying the continuous nature of the scales of significance and confidence as well as being a better basis on which to synthesize the results of repeated studies. Formal meta-analysis of a larger set of results of this sort would produce a consistent (and statistically significant) picture at odds with the misleading impression of inconsistency given by the significance tests and P-values.

10.6 Summary: Statistics

Many statisticians now recommend scientists curb their inappropriate overuse of null hypothesis significance testing and focus more on working with estimates and intervals. In order to do this is it necessary to understand the different types of intervals that exist and how they relate to each other. Working with estimates and intervals can help keep us closer to the biology we are interested in. While intervals have some of the same disadvantages as significance tests, they can help avoid dichotomous thinking of results as either statistically significant or non-significant. Presenting estimates and intervals is also essential for meta-analysis, one of our main tools for identifying repeatable, general results.

10.7 Summary: R

- The arm package coefplot() function is particularly useful when analysing differences between matched data.

10.8 References

Cumming, G. (2013) The new statistics: A how-to guide. *Australian Psychologist* 48: 161–70.

Wickham, H. (2016) *ggplot2: Elegant Graphics for Data Analysis*, second edition. Springer.

Analysis of Variance

11.1 ANOVA tables

Hopefully it is clear by now that a back-to-basics part of the 'new statistics' is a shift to an estimation-based approach that keeps us closer to the biology of interest. It is relatively straightforward to apply for simple situations like Darwin's maize and the Janka timber hardness data. However, as designs become more complicated the number of comparisons can become overwhelming, and working with estimates and intervals alone can become harder. For example, as designs get more elaborate the number of potential pairwise t-tests increases rapidly, as does the risk of false positive results (the more tests we make, the more likely we are to get false positives). It is then often useful to employ a complementary approach that can ask whether there is support for any differences among means at all, before diving into the detail of pairwise comparisons. Ronald Fisher invented such an approach with analysis of variance and the ANOVA tables used to present it. Although ANOVA tables tend to be associated primarily with analyses with categorical explanatory variables (factors), these tables can be generalized to analyses that include continuous explanatory variables (it may be more useful to think of them as linear-model analysis

The New Statistics with R: An Introduction for Biologists. Second Edition. Andy Hector, Oxford University Press. © Andy Hector 2021. DOI: 10.1093/oso/9780198798170.003.0011

tables). While ANOVA comes into its own with complex data sets, for teaching purposes we will start by applying it to Darwin's maize data.

11.1.1 R PACKAGES

```
library(arm)
library(ggplot2)
library(SemiPar)
library(SMPracticals)
```

11.2 ANOVA tables: Darwin's maize

Our simple linear model for Darwin's maize data from Chapter 6 was.

```
ls1 <- lm(height ~ type, data = darwin)
```

Because we only have one explanatory variable in this example, the model is called a one-way ANOVA. The general strategy of ANOVA is to quantify the overall variability in the data set, and then to divide it into the variability between and within groups (self- and cross-pollinated) and to calculate a signal-to-noise ratio. The greater the ratio of the signal to the noise, the more confident we can be that we have detected a real effect. The variability is quantified using a method called least squares. Least squares first quantifies the overall variability (the total sum of squares, SST) by measuring the differences from the individual data points to a reference point, or 'intercept' (Fig. 11.1(a)); the most intuitive one is the overall grand mean, but any value will do (and, as Box 11.1 explains, statistical software usually chooses a different benchmark).

Next, least squares quantifies how much of this overall variability is explained by classifying the data points into the treatment groups (Fig. 11.1(b)). This leaves the residual unexplained variation (Fig. 11.1(c)).

In this way, least squares quantifies the overall variability and splits it into signal and noise. As explained below, least squares works with the sums of the squared distances between individual data points and the

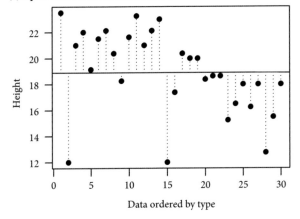

(a) Square and sum differences for SST

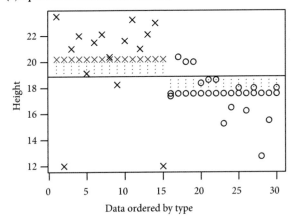

(b) Square and sum differences for SSA

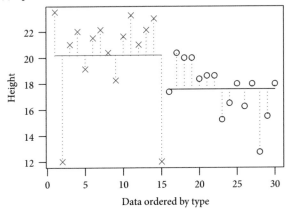

(c) Square and sum differences for SSE

Figure 11.1 Visualization of how the least squares method quantifies (a) total, (b) treatment, and (c) error sums of squares (SST, SSA, and SSE). The vertical lines measure the differences that are then squared and summed. The SST is calculated by measuring the differences from each data point to a reference point—the overall mean is the most intuitive (horizontal line). The vertical differences for the SSA are between the treatment-level means (the horizontal lines of 'fitted values' shown by the crosses and circles) and the overall mean. The differences for the SSE (right) are between the observed data values and the treatment-level means.

means—the 'sum of the squared differences', or 'sum of squares' (SS) for short. As the name implies, ANOVA works instead with the variances (which we have met in earlier chapters).

Box 11.1 - The linear-model 'intercept'

Explanations of how sums of squares are calculated usually measure the variability in the data relative to the grand mean, as this is the most intuitive choice. Indeed, our 'empty' ls0 model from earlier did estimate the grand mean as an intercept. However, the grand mean isn't usually of interest in itself, and the degree of freedom it costs to estimate the grand mean can be seen as wasted. However, we *are* usually interested in the treatment-level means (and the difference between them). Linear-model software is therefore efficient in using one of the treatment-level means as the 'intercept'. As we saw earlier, by default R applies alphanumeric rules to choose which factor-level mean is taken as the intercept. Most software works in a similar away (although there are more complex alternatives to this scheme of 'treatment contrasts' which are the default in some other software packages—but those are beyond the scope of this book).

The anova() function draws up an analysis-of-variance table:

```
anova (ls1)
```

```
## Analysis of Variance Table
##
## Response: height
##              Df   Sum Sq Mean Sq F value   Pr(>F)
## type          1   51.352  51.352  5.9395 0.02141 *
## Residuals    28  242.083   8.646
## ---
## Signif. codes:
## 0 '***' 0.001 '**' 0.01 '*' 0.05 '.' 0.1 ' ' 1
```

As wonderful a thing as the ANOVA table is, notice that the content is primarily about the statistical analysis and there is little biological information that we can directly extract. Contrast this with the table of coefficients (produced by the display() or summary() function) that we have mainly worked with to this point, which does contain immediate information about the biology (the average height of the selfed plants and the average difference in heights). This cautions that the ANOVA tables should be of secondary interest to us (it is amazing how many papers contain multiple ANOVA tables but are lacking basic summaries of the biological measurements). Nevertheless, ANOVA tables can be a really useful part of linear-model analyses—especially complex ones—so long as we use them appropriately. The ANOVA table has three rows (counting the headings) and six columns. The first (unnamed) column gives the source of the variation, in this case due either to the pollination treatment (signal) or to the residual unexplained variation (noise). The second gives the number of so-called *degrees of freedom* (DF), which is essentially an indication of the number of treatment levels (groups) and experimental units (the sample size), but with a twist—see Box 11.2.

Box 11.2 - Degrees of freedom

If I told you I had three numbers in mind that added up to 10, could you tell me the value of any of those numbers? No: they could take any of an infinite combination of positive or negative values so long as they summed to 10. If I tell you one of the numbers is 5, you are no better off: the other two numbers can still take any combination of values so long as they sum to $10 - 5 = 5$. However, if I tell you the second number has the value of 3 then you can tell me that the value of the last number must be 2. In other words, given a list of numbers and their sum total, all of the numbers are *free* to take any value except the last. The value of the last number is fixed by the sum total. Hence, in general, the total DF is $N - 1$ (in this case $3 - 1 = 2$). When referring to a treatment, the number of degrees of freedom is the number of treatment levels minus one, and for the residual it is the sample size minus one (and minus the DF taken by the treatment levels).

The third column gives the sums of squares. Unlike some statistical software packages, R does not give a final row for the total variation. If we want the total variation, we can simply add up the sums of squares for the signal and noise or take them from the anova() function output for the ls0 model given in Chapter 6 (note that only the sums of squares and degrees of freedom are additive, and not the other columns).

We have already met the quantity given in the fourth column, albeit under another name. The mean square is the variance we already know from earlier chapters. On the one hand, it is potentially confusing to have more than one name for the variance. However, at least 'mean square' is a name that gives an indication of how this quantity is calculated, which is by dividing the sums-of-squares value (for the row in question) by the number of degrees of freedom. Essentially, it gives an average amount of variability per treatment level or per experimental unit (but again with the 'N minus one twist'—strictly, it is an average per degree of freedom; see Box 11.2).

We shall look at how we decompose the total variation into signal and noise in a moment, but notice first that there is a single residual mean square (variance) that is calculated by pooling the variation within both samples in order to get a better estimate of the noise than would be the case if we calculated the variance separately for both groups based on the smaller sample size (as we did when first exploring this data set). This is particularly valuable when we have a greater number of treatments with smaller sample sizes, and was one of the key ideas in the development of ANOVA by Fisher. As far as I am aware, ANOVA—and linear models generally—always use the pooled estimate of the variance. Because of the use of a pooled estimate of the variance, the standard errors for estimates of the means of different groups can differ only if they have different sample sizes. If the sample sizes for different factor levels are the same then the numerator and denominator in the formula for the SEM are the same for all groups (take a look back at the formula for the SEM if this isn't clear).

11.3 Hypothesis testing: *F*-values

Returning to the table, we now have the variances (mean squares, MS) that the table and the analysis are named after—so how do we use them to perform ANOVA? The variance for the experimental treatment (pollination type) is our signal, while the unexplained residual variance is the noise. The signal-to-noise ratio is calculated by dividing the treatment variance by the residual error variance to produce the *F*-value (originally called the variance ratio but later renamed after Fisher, the inventor of ANOVA), given in the fifth column:

$$F = \frac{\text{MS}_{\text{treatment}}}{\text{MS}_{\text{residual}}}$$

In this case, the value of 5.9 means the estimated signal is nearly six times larger than the estimated noise. In a perfect world, if there were no signal then the treatment variance would be nothing more than another estimate of the noise—the error variance. If we could estimate both the signal and the noise perfectly, then if there was absolutely no effect of the treatment we would be dividing noise by noise and the variance ratio produced would be equal to one (the null expectation). In practice, we cannot get exact values for the signal and noise, since we are estimating them from samples. This means that these estimates will sometimes be too high and sometimes too low, and consequently the treatment variance will sometimes be estimated (or, rather, underestimated) to be smaller than the error variance, and occasionally we will get F-ratios less than one ('negative variance components'). So long as we are confident there are no problems with our data and analysis, then we can do little more than write these off as sampling error. Of course, the variance ratio could also be bigger than one simply due to sampling error, but the larger the signal is relative to the noise, the more confident we can be that it is not a false positive. The probability value, P, in the final column of the ANOVA table quantifies this for us (Box 11.3). The probability value of 0.02 says that there is a 2% chance of observing a signal-to-noise ratio at least this large if the null hypothesis is true, i.e. if there is no effect of pollination type (and the assumptions of the analysis are met).

Box 11.3 - Mushy Ps: Frequentist probability values

Unfortunately, the definition of the P-value in a so-called frequentist analysis like ANOVA is quite complex and has several limitations that are widely criticized (especially by non-frequentist statisticians such as those who prefer Bayesian statistics or information criteria). The frequentist P-value is based on an imaginary series of repeated analyses. The P-value is the probability of having a value of the test statistic (an F-value) equal to or greater than the observed variance ratio if the null hypothesis were true—that is, if there were no treatment effect (signal). In this case, if Darwin's experiment had been repeated many times and if there were no effect of pollination type, then we would expect only 2% of the experiments to produce a result of this size or larger. In practice, if our experiment reveals a positive effect of pollination with a P-value of this size, it attains the conventional lowest level of confidence where we can reject the null hypothesis of no effect (we can never be absolutely sure, of course, and critics of P-values argue that $P = 0.05$ results in too many false positives and a minimum around $P < 0.004$ would be more fitting). However, in many areas of science, a probability of 5% is still usually taken as the cut-off point at which we declare a result to be statistically significant. What's so special about 5%? Absolutely nothing! It was simply a convention used by Fisher, which was widely taken up and which has stuck ever since despite its drawbacks. When using $P < 0.05$, we are imposing an arbitrary cut-off point on a continuum for the convenience of being able to dichotomize results into statistically significant or non-significant. To make matters worse, the probability depends on the sample size. All else being equal, the larger the sample, the lower the P-value (due to increased statistical power). This makes it hard to use P-values to compare the strength of results from different analyses when the sample sizes vary. One recent trend in statistics is therefore to focus less on P-values and the dichotomy between significant and non-significant and instead to take a more continuous view by looking at estimates and confidence intervals. For this reason, from now on we will override the default setting in the options() function to turn off the stars used in the R output to highlight the level of significance.

The probability value in the ANOVA table is calculated from the F-distribution. Just as with the t-test, the F-test takes the sample size into account, and the probability assigned to a given F-value depends on the number of degrees of freedom of both the signal and the noise. We can see this using the R pf() function to re-create the P-value in the ANOVA table:

```
pf(5.9395, 1, 28, lower.tail = FALSE)
```

```
## [1] 0.02141466
```

The first three arguments are the F-value and the number of degrees of freedom for the signal (numerator) and the noise (denominator). The last argument in the function is set to give us the probability of being in the tail of the distribution with an F-value equal to or greater than the observed value (5.9)—in this case about 2%.

You will often see *P*-values presented with no supporting information—'naked *P*-values'—but the reader cannot interpret these without knowing what test they come from (or at least what type of test statistics the *P*-value is attached to) and the number of degrees of freedom (we might interpret a $P < 0.05$ result very differently depending on whether the number of degrees of freedom was 10 or 10 000!). Because of this, it is important that when reporting results we don't just give probability values but present them together with the *F*-value and both numbers of degrees of freedom. The conventional way to report the results would be something like this:

> The height of the cross-pollinated plants was significantly greater than that of the self-pollinated plants ($F_{1,28} = 5.9$; $P = 0.02$).

However, even with all the supporting details in place, this way of reporting the results focuses on the statistical testing, not the biology. One recent suggestion is to report the exact *P*-value and dispense with the phrase 'statistically significant' entirely to avoid this arbitrary dichotomization of results (Wasserstein et al. 2019). The report also doesn't tell us the average heights of the crossed and selfed plants, nor how different they were, or even which group was larger. It is better to try and include these details along with the results of the statistical tests:

> The self-pollinated maize measured 17.5 (95% CI: 16.0–19.1) inches in height on average, while the cross-pollinated plants had a mean height of 20.2 (18.6–21.8) inches—a difference of 2.6 (0.4–4.8) inches (one-way ANOVA: $F_{1,28} = 5.9$; $P = 0.02$).

11.4 Two-way ANOVA

As I'm sure you've guessed, two-way ANOVA includes two explanatory variables. Usually these are treatments of interest, but since the plants in Darwin's maize experiment were paired we can stick with the same example by simply adding the factor for the plant pair to the linear model:

```
ls2 <- lm(height ~ type + pair, data = darwin)
```

We can now compare the tables for the one-way and two-way ANOVAs
(turning off the criticized significance stars, as we'll discuss in greater depth
later):

```
options(show.signif.stars = FALSE)
anova(ls1)
```

```
## Analysis of Variance Table
##
## Response: height
##             Df   Sum Sq Mean Sq F value  Pr(>F)
## type         1   51.352  51.352  5.9395 0.02141
## Residuals   28  242.083   8.646
```

```
anova(ls2)
```

```
## Analysis of Variance Table
##
## Response: height
##             Df   Sum Sq Mean Sq F value Pr(>F)
## type         1   51.352  51.352  4.6139 0.0497
## pair        14   86.264   6.162  0.5536 0.8597
## Residuals   14  155.820  11.130
```

Note how the degrees of freedom and sums of squares for the factor
'pair' were initially included in the residual error term of the simpler model.
It's easiest to see this with the degrees of freedom, where the initial value
for the residual, 28, is split into 14 for 'pair' $(15 - 1)$ and 14 for the new
residual $(28 - 14 = 14)$. One unsatisfying thing about Darwin's maize
experiment is that the pairing doesn't work as intended. If the pairing had
worked, then we would expect 'pair' to explain more variability than the
residual error (noise) and to have an F-ratio < 1 when compared with
it. And recall that if pairing did nothing (i.e. plants within pairs were as

variable as plants from different pairs) then we would expect an F-ratio of ~1. However, here 'pair' actually explains less variation than expected—the mean square for 'pair' is ~6, which when divided by the residual has an F-ratio of 0.55. When mean squares are smaller than the residual, it implies a 'negative variance component'. There are two explanations for this. It could simply be due to sampling variability (i.e. the 0.55 is just an underestimate). Alternatively, it implies a problem with the experimental design or procedure. Darwin's experiment is actually very impressive, given he conducted it several decades before the principles of experimental design were understood. However, Darwin did not use randomization, and that may explain why the pairing does not work.

We've deliberately started with a simple example focusing on the effect of one treatment with only two levels. For a straightforward comparison like this, we could dispense with the ANOVA table entirely and just present the appropriate estimates and confidence intervals. However, as we'll see in the next two chapters, ANOVA tables can sometimes be a useful preliminary step in the analysis of complex data sets, where we can first ask whether there is any evidence for any differences at all (sometimes called an 'ensemble' test) before diving straight into large numbers of unfocused pairwise comparisons (with the risk of false positives that an undirected approach brings). The important thing is not to get lost in the P-values—history suggests they exert a strange and powerful siren's call on us. One of the 'new statistics' messages of this book is to work primarily with the estimates and intervals whenever possible, with P-values playing a secondary, supporting role rather than hogging the limelight as they have done over recent decades.

11.5 Summary

Analysis-of-variance tables can be constructed for all linear-model analyses (with any mixture of continuous and categorical variables). The tables partition the overall variability into signal(s) and noise, which can be compared using F-tests (variance ratio tests). For complex data sets where many pairwise comparisons could be performed, an initial F-test can provide an initial check of whether there is evidence for any differences at all, and reduces the risk of overtesting and the false positives that can be generated.

11.6 Reference

Wasserstein, R.L., Schirm, A.L., & Lazar, N.A. (2019) Moving to a world beyond '$p < 0.05$'. *The American Statistician* 73, sup1: 1–19.

Factorial Designs

12.1 Introduction

So far, we have seen how to conduct analyses with either continuous explanatory variables (regression) or categorical ones (ANOVA, t-tests). What if we have an analysis with two (or more) explanatory variables where interactions are possible? This chapter gives an introduction to factorial designs that combines two categorical explanatory variables in a way that allows us to look at interactions between them.

12.1.1 R PACKAGES

```
library(arm)
library(cowplot)
library(ggplot2)
```

12.2 Factorial designs

Fully factorial designs test for interactions by applying two or more treatments (factors) in all possible combinations (fractional-factorial designs are more complex and feature only a subset). We can then estimate the

The New Statistics with R: An Introduction for Biologists. Second Edition. Andy Hector,
Oxford University Press. © Andy Hector 2021. DOI: 10.1093/oso/9780198798170.003.0012

effects of each treatment alone and in combination. The example data set (Box 12.1) contains two factors, each with two levels, and the fully factorial design contains all four possible treatment combinations.

Box 12.1 - Ecological background

The example data come from Hautier et al. (2009). We want to analyse the response of the net aboveground plant biomass production ('biomass') of grassland plots in relation to two resource addition treatments—addition of fertilizer (±F) to the soil and addition of light (±L) to the grassland understorey. If biomass production is limited by soil fertility, we would expect to see an increase following fertilization. Similarly, if production is limited by low levels of light (shading) in the grassland understorey, we would expect addition of light to increase biomass. However, it is possible that productivity could be co-limited. For example, adding fertilizer alone may increase biomass production but that, in turn, could increase shading in the understorey so that light becomes limiting there. If this is the case, when we add fertilizer and light in combination we would expect to see a greater increase in productivity than when either resource is added on its own. In other words, the fertilizer and light addition treatments may interact. In this scenario there would be a positive, or synergistic, interaction where the two resources have a greater effect in combination than alone (negative, or inhibitory, interactions occur where treatments reduce each other's effects).

The data are available as a text file called Data_Factorial.txt from http://www.plantecol.org/contemporary-analysis-for-ecology/. By default, the directory that contains the R Markdown file is taken as the working directory (which can otherwise be set via the RStudio menus, Session -> Set Working Directory), so the simplest way to work is to put the data file in the same place as the Rmd file—we can then load the data and take a look at the first few rows as follows (see Appendix 12a):

```
Factorial <- read.table("Data_Factorial.txt", header = TRUE)
head(Factorial)
```

```
##    fert_ light_   fl_ biomass
## 1    F-     L-  F-L-   254.2
## 2    F-     L-  F-L-   202.0
## 3    F-     L-  F-L-   392.4
## 4    F-     L-  F-L-   455.3
```

```
## 5      F-      L-  F-L-     359.1
## 6      F-      L-  F-L-     386.5
```

The fertilizer and light additions are both either applied or not applied, forming a *fully factorial* design in which all four combinations are present, producing these four treatment combinations:

- control (no added fertilizer or light: F−, L−);
- fertilizer (fertilizer only: F+, L−);
- light (light only: F−, L+);
- addition of both resources in combination (F+, L+).

The dataframe shows two alternative ways of indicating the design. As always, it is a good idea to use the str() and summary() functions to explore the structure of the data set:

```
str(Factorial)
```

```
## 'data.frame':      64 obs. of   4 variables:
##  $ fert_  : chr   "F-" "F-" "F-" "F-" ...
##  $ light_ : chr   "L-" "L-" "L-" "L-" ...
##  $ fl_    : chr   "F-L-" "F-L-" "F-L-" "F-L-" ...
##  $ biomass: num   254 202 392 455 359 ...
```

We have 64 observations of four variables. The last column is the response variable—the biomass. The remaining three columns provide two alternative ways of indicating the experimental design. Note that I have used an underscore at the end of the explanatory-variable headings to make it easier to distinguish the names of the factors from the names of the factor levels and make the output less of a blur of F's and L's. The column 'fl_' has four levels, indicated by using plus and minus signs to show whether fertilizer (F) and light (L) have been applied. This column treats the design as if it were a one-way analysis of variance (like the analysis of Darwin's maize data). In combination, the first two columns allow us to instead look at the design as a factorial. The columns 'fert_' and 'light_' each indicate whether the resource in question has been applied or not, again using plus and minus signs. The summary() function provides some summary statistics for the response variable 'biomass':

```
summary(Factorial)
```

```
##      fert_              light_             fl_
##  Length:64          Length:64          Length:64
##  Class :character   Class :character   Class :character
##  Mode  :character   Mode  :character   Mode  :character
##
##
##
##      biomass
##  Min.    :152.3
##  1st Qu.:370.1
##  Median :425.9
##  Mean    :441.6
##  3rd Qu.:517.2
##  Max.    :750.4
```

12.3 Comparing three or more groups

So far, we have only performed a linear-model analysis (of Darwin's maize data) that has a single factor with only two levels. The 'fl_' column in the present dataframe allows us to expand that to a single factor with more than two levels. As we build up to the full factorial analysis, we can also perform a linear-model analysis with two factors but *no interaction*. To begin with, let's keep things simple and just look at the untreated control and the fertilizer-only and light-only treatments so that there is no combined treatment and therefore no possibility of an interaction. As usual, in R there are many ways to take a subset of a dataframe; one that does not require downloading any

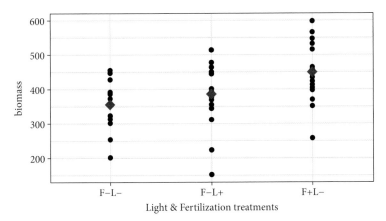

Figure 12.1 Individual plot biomass values (black points) with treatment means (red diamonds) for the data subset.

extra packages is to use the base-R subset() function as follows, to take the levels of the 'fl_' factor that are not equal (!=) to the coding for the combined treatment:

```
Sub <- subset(Factorial, fl_!="F+L+")
```

Plotting the data and superimposing the treatment means produces the graph shown in Fig. 12.1.

Let's begin the linear modelling where we left the Darwin's maize example and treat this as a 'one-way ANOVA' by using the single factor (which has only three levels in this subset of the data):

```
mod1 <- lm(biomass ~ fl_, data = Sub)
display(mod1)
```

```
## lm(formula = biomass ~ fl_, data = Sub)
##                 coef.est coef.se
## (Intercept)     355.79    21.41
## fl_F-L+          30.12    30.27
## fl_F+L-          93.69    30.27
## ---
```

```
## n = 48, k = 3
## residual sd = 85.63, R-Squared = 0.18
```

We can approach the display() output exactly as we did for Darwin's maize data except that we now have a third factor level. That means the output shows us one mean and the difference between this 'Intercept' and the means of the other two factor levels. By elimination, we can see that the control treatment (F−, L−) is taken as the intercept, and has a mean biomass of approximately 356 g with a standard error of 21 g. The light addition treatment has a difference in mean biomass from the control of +30 g, which is of a similar size to the standard error for this difference. In contrast, the mean for the fertilizer treatment is 94 g higher than the control, which is around three times as large as the SED. We can visualize the treatment differences using coefplot() (Fig. 12.2):

```
Fig12_2 <- coefplot(mod1)
```

In summary, while adding light alone produces a slight increase in the estimated mean biomass, a 95% confidence interval would include zero (and negative values down to around −30 g too), so we don't have any real confidence in light having a consistent effect. In contrast, fertilizer produces

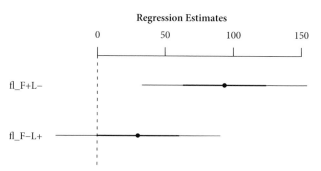

Figure 12.2 Effects of fertilizer (F+L−) and light (F−L+) addition on biomass relative to untreated controls (differences ±1 and 2 SEDs).

a substantial increase in yield, which we can have confidence in relative to the standard error of the difference.

12.4 Two-way ANOVA (no interaction)

In passing, as we build towards the full factorial analysis, we can fit a model with the two factors 'light_' and 'fert_'. The model below does not allow for an interaction yet (because the subset of the data excludes the combined treatment, it is not possible to look for one anyway), and the coefficients are identical to those in the model above:

```
display(lm(biomass ~ light_ + fert_, data = Sub))
```

```
## lm(formula = biomass ~ light_ + fert_, data = Sub)
##                 coef.est coef.se
## (Intercept)   355.79     21.41
## light_L+       30.12     30.27
## fert_F+        93.69     30.27
## ---
## n = 48, k = 3
## residual sd = 85.63, R-Squared = 0.18
```

Notice here that we have made only two pairwise comparisons: the light and fertilizer treatments have both been separately compared with the control. However, there is a third comparison we could make—between the light and fertilizer treatments. This raises a thorny issue: there are more possible pairwise comparisons than we have degrees of freedom (in this subset of the data, the three treatment levels give $3 - 1 = 2$ DF). The issue becomes increasingly problematic as the number of treatments and the number of resulting potential comparisons increase. One response is to throw caution to the wind, compare everything with everything, and then use one of the many alternative adjustments for multiple comparisons to take this into account. However, this can of worms is perhaps best left unopened whenever possible (Box 12.2).

Box 12.2 - Multiple comparisons: A fishy business

Like many things in statistics, adjustments for multiple testing are a topic where opinions differ widely. This is partly because there are situations where it is clearly important to correct for making multiple comparisons, but such adjustments are also overused in contexts where it would be better to avoid the need for multiple comparisons in the first place. For example, in a 'big data' analysis very many comparisons may have been made but it may be unclear how many. We expect about 1 in 20 tests at the $P < 0.05$ level to be significant by chance alone, so if we have performed many comparisons we would expect many false positive results. If this fact is not appreciated, or if the number of comparisons is not even reported, then we can end up misled by these false positive significant results. This was illustrated in an Ig Nobel Prize-winning piece of research that showed a test subject photographs of people displaying a variety of emotions and used MRI to demonstrate that different parts of the brain showed significant activity in response to the different facial expressions. The problem was that the test subject was a dead salmon! This potentially misleading result came about because the brain was divided into thousands of different parts and the significant responses were just the inevitable false positives. So, correcting for multiple comparisons is essential in many situations. On the other hand, many well-respected statisticians are critical of their unnecessary overuse, especially with the analysis of relatively simple designed experiments and data sets that are not 'big' (the focus of this book). For example, in their book Gelman & Hill (2006) argue that:

> We almost never expect any of our 'point null hypotheses' to be true. If we examine 100 parameters or comparisons we expect about 5% of the 95% intervals to exclude the true values. There is no need to correct for the multiplicity of tests if we accept that they will be mistaken on occasion.

So, if the number of comparisons is clear and we take this into account when interpreting the results we can often avoid adjustments for multiple comparisons (we adjust our expectations and inferences, not the statistics). One procedure that seems to be heavily overused is the automatic use of corrections for multiple comparisons followed by labelling of bar graphs with letters to indicate significant differences between means (means that cannot be distinguished statistically are coded with the same letter). This sounds useful but often indicates an unfocused approach and an inappropriate overuse of significance tests. The letter labelling is often redundant anyway, since the letters distinguish means that are visually different based on the error bars. Where differences are less clear, these post hoc pairwise tests often do not have the power to distinguish between means, resulting in many bars being labelled with overlapping combinations of letters, making for a graph cluttered with 'chart junk'. In other words, adding letters to graphs works best when you don't need them—when there are differences among means that are usually obvious from a comparison of appropriate error bars, so long as we understand the bars and know how to interpret them. A graph with error bars also gives a more continuous indication of the differences and variation in the data and helps avoid overuse of the dichotomous approach of significance tests. Instead, we can minimize concerns about multiple comparisons by keeping our analyses as simple and focused as possible and by performing the minimum number of essential hypothesis tests. However, when the number of comparisons is very large, unclear, or both, you will need to resort to the large literature on the various approaches to adjustment for multiple comparisons. When you need them, the multcomp package—and the supporting book by Bretz et al. (2010)—is a great resource.

Instead, with a little forethought, can we achieve our goals using only the two DF available to us? In this case, our main interest is in comparing the fertilizer and light treatments with the control. Notice that when this is the case the default organization of the summary() and display() functions is exactly what we want (if it is not, we know how to use the relevel() function to override the default organization and set the treatment level of our choice as the intercept). We could compare the other two treatments, but we don't have any particular interest in doing so, so let us resist the temptation to go fishing for significant results: the more comparisons, we make the more spuriously significant results we will get. In other situations, it may not be a case of a series of control-versus-treatment pairwise comparisons. Nevertheless, with a little thought beforehand, we may be able to specify a set of *a priori* comparisons (contrasts) that ask our main questions while staying within the limits of the degrees of freedom available and avoiding the complex and often unsatisfying adjustments for multiple comparisons.

To summarize, we have initially taken a subset of the data to examine a comparison of three or more treatments. The situation is relatively straightforward when we want to make a series of pairwise comparisons between a control and other manipulated treatments (as in this case). If we set the control treatment as the intercept then the default layout of the R output gives us the estimates and intervals that we need. However, our goal is to analyse the fully factorial design *including potential interaction effects.* Before we can return to the analysis of the full data set, we need a strategy that will allow us to estimate the degree of interaction.

12.5 Additive treatment effects

To get at the interactive effect, we need to be able to compare the effect of the combined fertilizer-plus-light treatment with what we would expect to happen if there were no interaction at all. Factorial ANOVA creates this no-interaction scenario by assuming that treatment effects are independent and therefore additive. In other words, if one treatment has an effect size of *A* and a second treatment has an effect size of *B* then when we apply the treatments in combination the linear-model ANOVA predicts the result to be *A* + *B*. What would this mean for the analysis of this example data set? Well, adding light on its own increased biomass by about 30 g and adding fertilizer on its own increased biomass by 94 g, so if their effects were independent and additive we would expect the combined treatment to produce about 124 g (30 + 94) more than the untreated control:

```
coef(mod1)[2] + coef(mod1)[3]
```

```
##   fl_F-L+
## 123.8187
```

Including the mean biomass for the control treatment gives us the expected means of the other two treatments (note, that although it is less visually immediate than working with the displayed values, it is better to use the square indexing brackets to work with the values of the coefficients directly to avoid rounding errors; so, instead of typing 355.79 + 30.12 + 94.69, we use the following code):

```
coef(mod1)[1] + coef(mod1)[2] + coef(mod1)[3]
```

```
## (Intercept)
##    479.6125
```

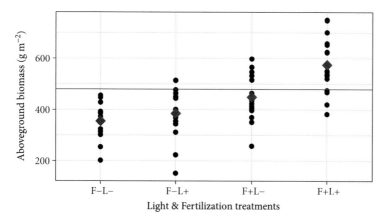

Figure 12.3 Individual plot biomass values (black points) with treatment means (red diamonds) for the fully factorial design (the horizontal blue line shows the additive expectation for the combined treatment).

Now that we have our additive prediction ($356 + 30 + 94 = 480$ g), it is time for the big reveal—we can plot the whole data set, including the fourth, combined treatment, and see how its mean compares with this additive prediction (shown as a horizontal line) (Fig. 12.3):

The mean biomass for the combined treatment is well above the additive prediction, suggesting that there may be a positive interaction. But how confident can we be in this effect? As usual, to answer that we have to judge the difference between the observed mean and the expected value relative to the background noise—the within-treatment variability. And for that we need the linear model for the full data set. To see its shortcomings, let's first do this as a 'one-way ANOVA' using the single factor with the four treatment levels:

```
mod2 <- lm(biomass ~ fl_, data = Factorial)
display(mod2)
```

```
## lm(formula = biomass ~ fl_, data = Factorial)
##
##               coef.est coef.se
## (Intercept) 355.79      23.14
## fl_F-L+      30.12      32.72
```

```
## fl_F+L-       93.69      32.72
## fl_F+L+      219.23      32.72
## ---
## n = 64, k = 4
## residual sd = 92.56, R-Squared = 0.47
```

Note that the point estimates are the same as for the three-level subset of the data but that adding back the data for the fourth treatment combination changes the pooled estimate of the variance (the residual error) and therefore the standard errors (compare the residual variance in the ANOVA tables for the two models to see this). Understanding the coefficients is key, so let's run through them again:

- The 'Intercept' row gives the mean biomass for the untreated control treatment with the standard error of the mean: 355.79 ± 23.14.
- The F−L+ row shows the *difference* between the untreated control and the light-only treatment with the standard error of the difference: 30.12 ± 32.72.
- The F+L− row gives the *difference* between the untreated control and the fertilizer-only treatment: 93.69 ± 32.72.
- The final F+L+ row presents the *difference* between the untreated control and the combined treatment: 219.23 ± 32.72.

Using coefplot() to visualize these coefficients produces the graph in Fig. 12.4 of the differences of the three manipulated treatment combinations relative to the control (note that the interaction is given first at the top of the graph):

```
Fig12_4 <- coefplot(mod2)
```

So, there is no detectable effect (relative to the noise) of adding light. In contrast, adding fertilizer alone has an effect (with the same SED) which is three times as large. We can be more confident that there is a positive effect of fertilizer because all the values in the confidence interval are clearly

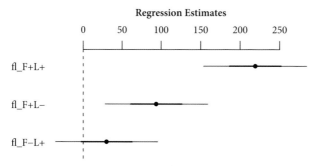

Figure 12.4 Effects of fertilizer (F+) and light (L+) addition and their combination on biomass relative to untreated controls (differences ±1 and 2 SEDs).

positive (running the confint() function for this model will show you that the lower bound of the confidence interval is around 28 g). Finally, for the same reasons we can be confident that the combined treatment has a clear positive effect—the increase in biomass is even greater (with the same confidence interval).

This is a good example of how an 'estimation-based approach' with confidence intervals can be much better than significance testing using *P*-values (and why many statisticians are encouraging us to use confidence intervals more and *P*-values less). This is a more sophisticated way of using confidence intervals than just asking whether zero is in the interval or not (which has the same limitations as using *P*-values in significance tests to declare results as 'statistically significant' or not). Instead, the confidence interval shows us the whole range of plausible effect sizes consistent with the data. Better still, the coefplot() function shows two intervals (by default ~68% and ~95%, i.e. ±1 and 2 SEs) rather than using just one level of confidence.

So, taking this all together, does it look like the additions of fertilizer and light interact? The combined treatment produces much more than either resource applied on its own, suggesting that light and fertilizer may indeed interact. Specifically, on their own light and fertilizer increase biomass by about 30 and 94 g, respectively, leading us to expect a 124 g increase when they are applied in combination if they do not interact. Instead, adding the two resources together increases biomass by around 220 g—nearly 100 g more than the additive prediction. So, with a bit of simple arithmetic we can use the one-way ANOVA approach to estimate the size of the interaction (at nearly 100 g more than expected). But, because we have to calculate this estimate ourselves, we have no standard error to go with it. To really get at the interaction we need a linear model that takes account of the factorial design.

12.6 Interactions: Factorial ANOVA

In contrast to the one-way ANOVA approach, the factorial ANOVA allows us to estimate the interactive effect directly along with a standard error, which we can use to calculate a measure of confidence (and to formally test the interaction if we want). Before we get to the formal statistical modelling, it is useful to visualize the interaction. R has a handy interaction.plot() function doing just that (unfortunately, because the interaction.plot() function lacks the 'data' argument we have to use the with() function to tell R which dataframe to use) (Fig. 12.5(a)):

```
Fig12_5a <- with(data = Factorial,
  interaction.plot(light_, fert_, biomass))
```

We can put light on the x-axis and have different line types for fertilization, or the other way around (Fig. 12.5(b)):

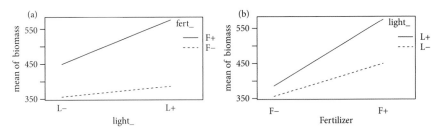

Figure 12.5 Interaction plot with (a) light addition on the x-axis; (b) fertilizer addition on the x-axis.

```
Fig12_5b <- with(data = Factorial,
    interaction.plot(fert_, light_, biomass, xlab = "Fertilizer"))
```

12.6.1 FACTORIAL ANOVA IN R

To perform the factorial ANOVA, we need to use the separate factors for fertilizer and light rather than the single 'fl_' factor. The key change is that we alter the model formula so that we ask for the effect of light, plus the effect of fertilizer, plus any interactive effect of the two ('fert:light'). A shorthand way to write the model (ignoring the underscores) is fert*light, which expands out to fert + light + fert:light:

```
mod3 <- lm(biomass ~ light_ + fert_ + fert_:light_, data = Factorial)
display(mod3)
```

```
## lm(formula = biomass ~ light_ + fert_ + fert_:light_, data = Factorial)
##                    coef.est coef.se
## (Intercept)         355.79    23.14
## light_L+             30.13    32.72
## fert_F+              93.69    32.72
## light_L+:fert_F+     95.41    46.28
## ---
## n = 64, k = 4
## residual sd = 92.56, R-Squared = 0.47
```

Notice how the factorial analysis has less power to estimate the interaction, leading to a larger standard error than those for the main effects (where the data set is divided in half to estimate each main effect while averaging over the levels of the other factor, rather than into quarters). We can visualize the table of coefficients—the differences from the intercept at least—using the coefplot() function (Fig. 12.6):

```
Fig12_6 <- coefplot(mod3)
```

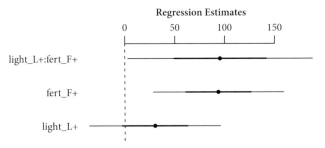

Figure 12.6 The coefplot() graph for the factorial model.

When we use the factorial combination of the two treatments, the last line of the table of coefficients estimates the size of the interaction (i.e. the non-additive effect) at around 95 g. Just to be clear, if the combined treatment produced a biomass equal to the additive prediction (the biomass of the control plus the increase due to fertilizer plus the increase due to light) then the estimate of the interaction effect would be zero instead of the 95 g observed here. The linear model that used the single factor 'fl_' compared the combined treatment with the untreated control and gave a difference of 219 g. The factorial model shows that this is made up of the additive effects for the light and fertilizer additions plus the interactive effect (30 + 94 + 95).

To summarize, the full table of coefficients for the factorial model can be used as follows. Apart from the last row, everything is the same as explained above and as in earlier chapters for models without interactions. The intercept is the mean for the unmanipulated control. Adding the second coefficient to the first gives the mean for the light-only treatment:

```
coef(mod3)[1] + coef(mod3)[2]
```

```
## (Intercept)
##    385.9188
```

Adding the third coefficient to the first gives the mean for the fertilizer-only treatment:

```
coef(mod3)[1] + coef(mod3)[3]
```

```
## (Intercept)
##    449.4875
```

The new part is the combined treatment. To get the estimate of the mean for the final treatment, we take the baseline intercept (which receives neither fertilizer nor extra light in the understorey) plus both main effects, plus the final interaction term:

```
coef(mod3)[1] + coef(mod3)[2] + coef(mod3)[3] + coef(mod3)[4]
```

```
## (Intercept)
##    575.0188
```

The table of coefficients also gives a column of standard errors. Each standard error corresponds to the estimate (a mean or a difference between two means) given in the column to its left (which we have already worked out). The first row gives a mean, so the standard error is the SEM. The next two rows give differences between means, so the standard errors are SEDs. The only new thing is the final row, which reports the interaction—the difference between the additive prediction and the estimate of the mean for the combined treatment with a standard error for this difference (an SED). As usual, confidence intervals for these estimates can be extracted using the confint() function:

```
confint(mod3)
```

```
##                       2.5 %      97.5 %
## (Intercept)      309.508816 402.07868
## light_L+         -35.331781  95.58178
## fert_F+           28.236969 159.15053
## light_L+:fert_F+   2.836382 187.97612
```

Don't make the mistake of looking at the row for the main effect of the light treatment and reporting that there is no evidence for an effect of light because zero is included in the interval. Remember that the main effect of light gives the average effect of light across the two fertilizer treatments. However, because of a detectable interaction, this average effect does not tell the full story—whether light has an effect depends on the level of fertilization, and the effect of adding light to the understorey is stronger than expected for fertilized plots. In other words, start at the bottom of the table of coefficients (and the ANOVA table too) and work upwards from interactions to main effects. If you detect evidence for an interaction then all treatments involved in that interaction need to be considered *even if the tests for their average main effects are non-significant*. In some cases, main effects may be much stronger than interactions and you may want to report both the average effects and the (weaker) interactive effect. Use the scientific context to judge what is *biologically* significant.

What about ANOVA tables for factorial models? ANOVA tables present a single summary of a whole analysis. This is useful in many ways, at least in simple situations. However, one disadvantage is that the sequential nature of ANOVA is not well reflected in the tables (as we have seen, it's better to start at the bottom and work up). This is not such an issue with data sets like this example that are balanced and orthogonal so that the effects of the two treatments can be estimated independently of each other (you get the same coefficients if you type Light + Fert or Fert + Light), but it is more of an issue in the analysis of unbalanced data sets, as we'll explore in Chapter 14. However, even in balanced cases ANOVA tables can potentially lead unwary readers astray in cases like this where interactions are significant but some main effects are not. Rather than getting a single summary of the analysis, we can work through it in sequence by using the anova() function to compare pairs of models, dropping one term at a time. First, we need a model without the interaction:

```
mod4 <- lm(biomass ~ light_ + fert_, data = Factorial)
```

Which we can then compare with the model with the interaction term using the anova() function with the pair of models as two arguments (remembering to turn off those attention-grabbing stars):

```
options(show.signif.stars = FALSE)
anova(mod3, mod4)
```

```
## Analysis of Variance Table
##
## Model 1: biomass ~ light_ + fert_ + fert_:light_
## Model 2: biomass ~ light_ + fert_
##   Res.Df    RSS Df Sum of Sq      F  Pr(>F)
## 1     60 513998
## 2     61 550407 -1    -36409 4.2501 0.04359
```

R renames the pair of models 1 (complex) and 2 (simpler) and compares them using the same *F*-test as reported in the ANOVA table, but this sequential approach guides us through the analysis in a logical order. The lower part of the output gives, from left to right, the identity of the model (1 or 2), the number of degrees of freedom, and the *residual* sum of squares. Next comes the change in DF (−1) and in the SS and the probability of observing such a change (or a more extreme one) if the null hypothesis were true and there was no interaction. The significance test rejects the null hypothesis of no interaction effect (although only just). The drop1() function does a similar thing:

```
drop1(mod3, test = "F")
```

```
## Single term deletions
##
## Model:
## biomass ~ light_ + fert_ + fert_:light_
##               Df Sum of Sq    RSS    AIC F value  Pr(>F)
## <none>                     513998 583.43
## light_:fert_   1     36409 550407 585.81  4.2501 0.04359
```

The advantage of the drop1() function is that it starts with the (highest-order) interaction, and if that is significant it stops there (rather than inappropriately continuing on to look at the main effects in the presence of an interaction as a standard ANOVA table does). The drop1() function also provides the value of the Akaike information criterion (AIC), which can be used for model selection as an alternative to P-values, as we'll explore in more detail later.

If you are going to perform significance tests, it is important not to report just the 'naked' P-values but to accompany them with the essential supporting information—what analysis was performed, the value of the test statistic (t, F, etc.), the exact value of P (not just $P < 0.05$), and the number of degrees of freedom (or sample size), something like

> There was an interactive effect of light addition and fertilization (factorial ANOVA: $F_{1,60} = 4.3$, $P = 0.044$) in which the combined treatment produced substantially more biomass (95.4 g [95% CI: 2.8, 188]) than expected when the two resources were added alone.

However, don't make the mistake of focusing on the P-values and forgetting to present the biologically more interesting effect sizes (with their confidence intervals). Providing estimates and intervals is usually preferable, as it allows a test of the interaction but also provides information on the effect size (how strong the interaction is in terms of the effect on biomass production). As we've seen, in many cases it may be sufficient to present the estimates and confidence intervals alone and dispense with the P-values (or to present them as supplementary material when needed). We could make further pairs of models to test the average main effects of light and fertilizer, but since the interaction tells us that both are important (with the effect of each dependent on the other) they could be misleading—the interaction provides the main result.

12.7 Summary: Statistics

This chapter has taken us from simple 'one-way' designs to more complex factorial designs and extended the simple linear model to include interactions as well as the average main effects. Interactions are assessed relative to an additive expectation where the treatment effects are independent of each other. Interactions can be positive, when effects are more than additive, or negative, when they are less than expected. When all four treatment combinations were treated as if they were four separate treatments, there was a stronger effect of fertilization on biomass production than of light addition to the understorey. However, this analysis could not explicitly address the interaction effect. Factorial analysis can address interactions, and in this case it supported a interactive effect: when understorey light and fertilizer addition were combined, the increase in biomass was greater than the additive expectation. This suggests that biomass production can be limited by different factors in the understorey and in the upper canopy layer. For example, following the addition of fertilizer, production in the understorey appears to become more strongly light limited, because when light is added to the understorey of fertilized plots the total aboveground biomass increases substantially.

12.8 Summary: R

- The subset() function provides a base-R way to take a subset of a dataframe.
- Interactions can be visualized using the interaction.plot() function.
- Pairs of models can be compared using the anova() function.

12.9 References

Gelman, A. & Hill, J. (2006) *Data Analysis Using Regression and Multi-level/Hierarchical Models.* Cambridge University Press.

Hautier, Y., Niklaus, P. A., & Hector, A. (2009) Competition for light causes plant biodiversity loss after eutrophication. *Science* 326: 636–38.

Bretz, F., Hothorn, T., & P. Westfall (2010) *Multiple Comparisons Using R.* Chapman and Hall/CRC.

Appendix 12a: Code for Fig. 12.3

Here is the ggplot2 code for the final figure for the full data set:

```
ylabel <-
  expression(paste("Aboveground biomass (g m"^"-2",")"))
Fig12_3s <-
  qplot(data = Factorial, x = fl_, y = biomass) +
  labs(x = "Light & fertilization treatments", y = ylabel) +
  theme(legend.position = "none") +
  geom_hline(yintercept = 480, colour = "blue") +
  stat_summary(fun = mean, geom = "point", colour = "red",
    shape = 18, size = 4) +
  theme_bw()
```

Analysis of Covariance

13.1 ANCOVA

The previous analysis dealt with an interaction between two categorical explanatory variables. We can also examine interactions between a factor and a continuous explanatory variable, an analysis that usually goes by the name of ANCOVA (although the terminology can be a bit confusing, as explained in Box 13.1).

13.1.1 R PACKAGES

```
library(arm)
library(ggplot2)
library(Sleuth3)
```

┤ **Box 13.1 - ANCOVA, covariates, and general linear models** ├

In the context of experiments, ANCOVA refers to a design with one factor and one continuous explanatory variable. One way to think of it is as a combination of the regression and one-way ANOVA linear models from earlier chapters. As we've seen, one advantage of R is that it takes a general approach and the linear-model function, lm(), can work with explanatory variables that are categorical factors ('ANOVA') or continuous ('regression') and with combinations of them ('general linear models'). One potentially confusing point is that general linear models are limited to using the

The New Statistics with R: An Introduction for Biologists. Second Edition. Andy Hector,
Oxford University Press. © Andy Hector 2021. DOI: 10.1093/oso/9780198798170.003.0013

normal distribution, whereas *generalized* linear models (GLMs or GLIMs) allow a broader range of distributions (as we'll see later in the book). By having just one factor and one continuous variable, ANCOVA can be seen as the simplest general linear model. However, ANCOVA is also used to describe analyses in which we need to adjust for the effects of uncontrolled variables when assessing the effects of the design variables. For example, imagine we were assessing the effects of an experimental treatment on the growth of a study organism but where the initial size varied at the start (see Chapter 14 for just such a case). Before assessing the effect of the experimental treatment on growth, we would need to adjust for differences in initial size. This type of analysis is also sometimes referred to as an ANCOVA with initial size as the covariate (although, just to make things even more confusing, some statisticians refer to all variables as covariates).

13.2 The agricultural pollution data

As an example of an ANCOVA, we will take a subset of the variables from an experimental study of the effects of low-level atmospheric pollutants and drought on agricultural yields. The aim of the experiment was to see how the yields of two varieties of soya bean, called Forrest and William, were affected by two pollutants (low-level ozone, O_3, and sulphur dioxide, SO_2) and how these pollutants interact with water stress. We will begin by looking at the effect of each of the pollutants with water stress in separate ANCOVAs before combining all three variables into a general linear model. We will only look at the William variety here, as it shows more complex responses (you can look at the Forrest variety yourself as an exercise). The dataframe, called case1402, is in the R package Sleuth3 (or the earlier Sleuth2), which accompanies the book *The Statistical Sleuth* (Ramsey & Schafer 2013):

```
library(Sleuth3)
head(case1402)
```

```
##           Stress  SO2   O3  Forrest  William
## 1 Well-watered  0.00 0.02     4376     5561
## 2 Well-watered  0.00 0.05     4544     5947
```

```
## 3 Well-watered 0.00 0.07    2806    4273
## 4 Well-watered 0.00 0.08    3339    3470
## 5 Well-watered 0.00 0.10    3320    3080
## 6 Well-watered 0.02 0.02    3747    5092
```

First, we have the three explanatory variables for water stress (Stress), sulphur dioxide (SO2), and low-level ozone (O3). The response is the yield of soya beans—here we have two versions, one for the variety William and one for Forrest. As usual, it is essential to understand the structure of the data set, so, as always, we run the str() and summary() functions:

```
str(case1402)
```

```
## 'data.frame':    30 obs. of  5 variables:
##  $ Stress  : Factor w/ 2 levels "Stressed","Well-watered":
##                2 2 2 2 2 2 2 2 2 ...
##  $ SO2     : num  0 0 0 0 0 0.02 0.02 0.02 0.02 0.02 ...
##  $ O3      : num  0.02 0.05 0.07 0.08 0.1 0.02 0.05 0.07 0.08 0.1 ...
##  $ Forrest : int  4376 4544 2806 3339 3320 3747 4570 4635 3613 3259 ...
##  $ William : int  5561 5947 4273 3470 3080 5092 4752 4232 2867 3106 ...
```

The dataframe has 30 rows and five columns.

```
summary(case1402)
```

```
##            Stress          SO2                 O3
##  Stressed    :15    Min.   :0.00000    Min.   :0.020
##  Well-watered:15    1st Qu.:0.00000    1st Qu.:0.050
##                     Median :0.02000    Median :0.070
##                     Mean   :0.02667    Mean   :0.064
##                     3rd Qu.:0.06000    3rd Qu.:0.080
##                     Max.   :0.06000    Max.   :0.100
##     Forrest         William
##  Min.   :2158    Min.   :2101
##  1st Qu.:3245    1st Qu.:2889
##  Median :3478    Median :3428
##  Mean   :3699    Mean   :3635
##  3rd Qu.:4327    3rd Qu.:4263
##  Max.   :5573    Max.   :5947
```

The first of the three explanatory variables is a two-level factor for drought stress—the soya beans were grown under well-watered control conditions or under drought stress. The soya bean plants were also exposed to controlled gradients of two atmospheric pollutants. Low-level ozone was applied at five levels and sulphur dioxide at three levels. Here we have a choice: we could treat these atmospheric pollutants as factors too, but, because they form points on a continuum, we are going to see if we can treat them as continuous explanatory variables (because, if it works, we will have a simpler linear relationship with yield).

By looking at the full data set and these summaries we can see that the experiment has a balanced $2 \times 3 \times 5$ fully factorial design (with respect to water stress $\times$ SO_2 $\times$ O_3). By 'balanced' we mean that it has equal numbers of replicates for each treatment combination. It has a completely randomized design (no blocking), in which treatment combinations were randomly allocated to one of 30 open-top growth chambers.

13.2.1 WARNING: R REGENERATES!

Just like Doctor Who, R periodically regenerates, in R's case rather more often than the time-travelling Gallifreyan—there is a new incarnation every six months. R packages must follow suit and can also have a life of their own, for example to match the edition of the book they accompany. The Sleuth package is in its third version, presumably tracking the editions of the book it accompanies. Checking the structure pays off with this data set, as one detail seems to change depending on which version of the package you are working with. As we know, R usually follows an alphanumeric approach when deciding which factor level (or combination of levels) will be taken as the 'intercept' in the linear model. With this data set, the Stress factor has two levels, Stressed and Well-watered, meaning that the Stressed level would normally be taken as the intercept in the analysis. However, in some previous versions of the package (including the one used in Chapter 7 of the first edition of *The New Statistics with R*) the package authors seem to have overridden the R default and instead made the Well-watered factor level the intercept. Presumably this was because you could argue that the well-watered treatment is the more usual situation (the control situation if you like) and therefore a more natural 'intercept'. Here, the str() function output lists the factor levels in alphabetical order—in this case the package authors seem to have gone with the R default, but since this changes with the package version we need to double-check this when interpreting the display() or summary() function output. This means there will be small differences in this analysis from the one reported in the first edition of this book (and materials based on that).

13.3 ANCOVA with water stress and low-level ozone

We are going to focus on ozone and water stress (ignoring SO_2 for now) and perform an analysis of covariance of their impacts on the yield of one of the varieties (William). In this type of experiment soya yields are normally analysed after automatic log transformation, so we will do the same for consistency (we need to keep this transformation in mind when interpreting the results, particularly the interaction, since multiplicative effects become additive on a log scale). We might expect both water stress and low-level ozone pollution to decrease yields. It also seems feasible that a plant exposed to either of these stresses might be more vulnerable to the effect of the other, and that we might find a synergistic interaction. On the other hand, there is a limit to how far yields can fall—we can't have negative values—and this may constrain things.

We can produce a panel plot of the effects of ozone on the yield of the William soya variety as follows. Before we start, because we are going to draw a few variations on this figure it is also a good idea to create objects for the axis labels so that we only have to write them out in full (or make any future changes) once:

```
xlabel <- expression(paste("Ozone (", mu, "L L"^"-1", ")"))
ylabel <- expression(paste("Log Yield (kg ha"^"-1", ")"))
```

In the qplot() code, the facets argument sets up a grid of panels by specifying *rows~columns* (in this case a point serves a placeholder to indicate there are no rows). Left to its own devices, ggplot2 will limit the plot to the space occupied by the data; here we can use xlim() to extends the *X*-axis back to zero, where the regression intercept is. There are a couple more things to add to the figure, so this preliminary version (zero) is not printed out (you know how to enter its name to see it if you want to):

```
Fig13_0 <- qplot(data = case1402, x = O3, y = log(William),
  facets = . ~ Stress, xlab = xlabel, ylab = ylabel) +
  xlim(0, 0.1)
```

We can visualize the ANCOVA as two regressions, one for each level of the water stress factor, using geom_smooth() to add the slopes (and, since we have two response variables in this data set, let's use ggtitle() to make it clear which we are looking at) (Fig. 13.1):

```
Fig13_1 <- Fig13_0 + geom_smooth(method = "lm") +
  ggtitle("Soya bean variety: William")
Fig13_1
```

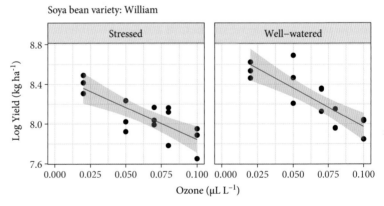

Figure 13.1 Yield (natural-log transformed) of William-variety soya beans as a function of low-level ozone and water stress.

Based on this graph, do you think there are effects of water stress and ozone on yield, and do you think they interact?

The most striking thing about the figure is the decline in yield with increasing ozone. The second most striking thing is that the intercept for the stressed treatment looks as if it might be reduced relative to the well-watered control. The third is that the slope for the stressed treatment is shallower than that for the control, but the difference appears to be small relative to the uncertainty. A model that considers the possibility of an interaction looks similar to the two-way factorial ANOVA model from the previous chapter, the only difference being that one of the explanatory variables is continuous:

```
w1 <- lm(log(William) ~ O3 * Stress, data = case1402)
```

Because the original researchers followed the convention in their field and automatically log-transformed the yields for analysis, we will skip the check of the residuals to save space (while they are not perfect, there are no major problems—try it) and we will move on to look at the estimates and intervals (as always, if you don't have the arm package display() function then use the base-R summary() instead):

```
display(w1)

## lm(formula = log(William) ~ O3 * Stress, data = case1402)
##                          coef.est coef.se
## (Intercept)               8.49     0.10
## O3                       -6.47     1.40
## StressWell-watered        0.26     0.14
## O3:StressWell-watered    -1.35     1.98
## ---
## n = 30, k = 4
## residual sd = 0.15, R-Squared = 0.71
```

As usual, the challenge is to work out what the rows of the table of coefficients show. Luckily, the table works in the same general way as for factorial ANOVA (we just have to think in terms of one factor and one continuous explanatory variable instead of two factors). If we think of this ANCOVA as regressions within the factor levels then it is clear that the table of coefficients will contain information on four quantities: two regression intercepts and two slopes (information on an intercept and slope for each factor level). As usual, the linear model will take one factor level as the 'intercept' and give us the intercept and slope for one of the regressions in the upper two rows, while the bottom two rows will give the *difference* in intercept and slope (as usual, we then have to calculate the intercept and slope for the other treatment level using these differences). Let's start at the bottom of the table of coefficients—the label for the last row makes it clear it involves both ozone and the stress treatment (it names the Well-watered level in this case), so this obviously addresses the interaction (we'll return to how in a moment). Moving on up, the label for the third row indicates that it addresses the effect of the well-watered treatment. That means, by elimination, that the first two rows must show the ozone effect for the Stressed watering treatment level (the R alphanumeric default has been applied here). Ozone is a continuous explanatory variable, so this is the regression part of our ANCOVA and the table must show the regression intercept and slope for the stressed water treatment. Now that the first two rows are deciphered, we can return to the effect of water stress (the factor). The third row gives the difference in intercepts for the regressions in the well-watered and stressed treatments. The last row addresses the interaction: the difference in slopes in the stressed and well-watered treatments. With some simple arithmetic, we now have the four values for the two intercepts and slopes, which we can reality-check versus the graph.

The question is, do we need different relationships for the different water stress treatments or does it look like there is a common relationship? As with the factorial ANOVA, we need to look at the final row of the table of coefficients, which addresses the interactive effect. The coefficient for the difference in slopes is −1.35 with a standard error of 1.98. The estimated difference is less than its standard error, so our conventional rule of thumb suggests no support for the interaction. If we wanted to be more exact, we could use confint() to see if the confidence interval for the interaction effect includes a value of zero (zero interaction) or not. We can also use the anova() function to more formally test the interaction:

```
anova(w1)
```

```
## Analysis of Variance Table
##
## Response: log(William)
##             Df  Sum Sq Mean Sq F value      Pr(>F)
## O3           1 1.13812 1.13812 51.9277 1.185e-07
## Stress       1 0.23764 0.23764 10.8426  0.002859
## O3:Stress    1 0.01013 0.01013  0.4621  0.502639
## Residuals   26 0.56985 0.02192
```

As we saw earlier, the drop1() function is smarter in 'respecting marginality' and first assessing the interaction (leaving the main effects out of the picture until it's clear whether it makes sense to look at them):

```
drop1(w1, test = "F")
```

```
## Single term deletions
##
## Model:
## log(William) ~ O3 * Stress
##               Df Sum of Sq     RSS     AIC F value Pr(>F)
## <none>                     0.56985 -110.91
## O3:Stress      1  0.010129 0.57998 -112.38  0.4621 0.5026
```

The first row of the drop1() output gives the residual sums of squares for the full model given above the table (as we can see if we look back at the ANOVA table). The second row reports the change in the degrees of freedom and sums of squares caused by dropping the interaction and reports an F-test for the deletion of this single term. It also gives values of the AIC, which can be used for model selection (Box 13.2).

Box 13.2 - Information criteria

The field of information theory provides an alternative basis for model comparison and selection through the use of information criteria that combine goodness of fit (the log-likelihood) and model complexity (the number of parameters). Unfortunately, there is a veritable alphabet soup of information criteria, but the original and most widely used is the AIC, or Akaike information criterion (and its version for small sample sizes, AICc):

$$AIC = -2\log - likelihood + 2\,k$$

where k is the number of estimated parameters (including estimated error terms when using maximum likelihood). We will meet likelihood again later in the context of generalized linear models. For now, we can note that -2 log-likelihood is the deviance—which, when using normal least squares, is the sum of squares we are already familiar with. In other words, the 2 can be seen as a scaling factor that we don't need to worry about—it seems Akaike put it in 'for historical reasons' to convert the log-likelihood to deviance (with its equivalence to the familiar sum of squares). But why convert the log-likelihood to deviance? Statisticians often work with the deviance because it approximately follows the chi-squared distribution, which can be used to derive levels of confidence and P-values (in the same way as we use the normal distribution). We don't need to worry about the minus sign either; it's also there for convenience, in this case because it is computationally easier to search for a minimum than a maximum. So the left-hand term of the equation is just a conventional measure of model fit that we are already familiar with. The right-hand term is often referred to as the penalty term (although some statisticians feel this is an oversimplification). The AIC adds twice the number of parameters. Again, the 2 can be seen as a scaling factor—it's just there to complement the doubling of the log-likelihood. If we cancel the 2s out, the formula simplifies to the negative log-likelihood (a measure of goodness of fit) plus the number of parameters (the penalty for complexity). We can view the value of the information criterion as a measure of efficiency (or parsimony) that balances the likelihood versus the number of parameters to find the model that is most efficient (parsimonious) in giving the best fit for the smallest cost. Because we are minimizing the likelihood, the smaller the negative log-likelihood, the better the model fit, and so we *add* the cost in terms of the number of parameters, and the most parsimonious model is the one with the *lowest* AIC value. This can be confusing, since with test statistics the larger the value, the more significant the term. With information criteria, because we use the negative log-likelihood as a measure of goodness of fit, the lower the information criterion value, the better the model (all else being equal). See Anderson (2008) for a primer on the AIC.

The F-ratio for the interaction provides no support for an interaction. The model comparison using the AIC is done as follows (technically it is not a test). As explained in Box 13.2, the AIC balances goodness of fit versus complexity to give a measure of parsimony for which the model with the lowest AIC value is the best, although models within two AIC units of each other are indistinguishable (all else being equal). In this case the model with the interaction included has an AIC of −110.91 and the model with the interaction removed has a value of −112.38. There is no information in the AIC values themselves, since their calculation involves arbitrary constants, so it is recommended to set the lowest AIC value to zero and report the *differences* in AIC values ('delta-AIC values') between models—in this case this difference is 1.47. At first glance, the rule of thumb that models within two AIC units of each other are indistinguishable might seem to suggest we could take either the model with the interaction or the one without, but this is not the case. Instead, think of it like this. Including the interaction in the model leads to no improvement, and so there is no support for the interaction term. We may have models with indistinguishable AIC values (within two units) and equal complexity, in which case we cannot distinguish between then. But here we have a pair of *nested* models, one simpler and one more complex, and when they have similar AIC values we should generally prefer the one with fewer parameters. In this case, that means the AIC and the *F*-test agree in preferring the simpler model (but significance tests and the AIC will not always agree—they are different approaches). We will look at model selection using information criteria again in Chapter 17.

13.4 Interactions in ANCOVA

We are now faced with a choice: should we keep the model as it is and move on to interpret the main effects (having shown there is no sign of an interaction) or should we follow the approach of model simplification and remove unimportant variables until we end up with a 'minimal adequate model' that includes only terms for which we have some support? There is no definitive answer, and you can read different advice in different places. Fashions have also changed over time, and while model simplification used to be generally recommended, some now advise retaining all terms (so long as they are clearly unimportant). However, this can lead to ambiguity, and whichever model you present it is safest to check whether retaining or dropping the interaction changes the result and interpretation of the analysis, as we'll see in Chapter 14. Model simplification does have the virtue of simplicity. Dropping the interaction here produces a model with the main effects of water stress and ozone—we still have two regressions, but now removing the interaction constrains them to share the same slope. The questions now are: is there a relationship (slope) with ozone, and is there a common relationship, or do we need two lines with different elevations for the two water stress treatments?

```
w2 <- lm(log(William) ~ O3 + Stress, data = case1402)
display(w2)
```

```
## lm(formula = log(William) ~ O3 + Stress, data = case1402)
##                      coef.est coef.se
## (Intercept)            8.53     0.07
## O3                    -7.14     0.98
## StressWell-watered     0.18     0.05
## ---
## n = 30, k = 3
## residual sd = 0.15, R-Squared = 0.70
```

An initial look at the estimates and standard errors for the slope with ozone (-7.14 ± 0.98) and the difference in intercepts (0.18 ± 0.05) makes it plain that we can be confident in both effects (which we could confirm more precisely using confidence intervals).

13.5 General linear models

This chapter has focused on ANCOVA of designed experiments. As noted above, because it combines one continuous and one categorical explanatory variable one way to think of ANCOVA is as the simplest form of a general linear model. By way of example, we will finish by going one step further to look at a more complex GLM analysis that combines water stress, ozone, and sulphur dioxide. One question is how complex we should make the model. The design allows us to include the three-way interaction, but we may not want to unless we have an a priori question that motivates its inclusion. With even more complex designs, the higher-order interactions may be effectively uninterpretable, arguing against their automatic inclusion. For these data we don't know what the *a priori* hypotheses were, but it seems reasonable that all three stresses could interact to further reduce yield, so we could try a model that includes all three explanatory variables and their interactions:

```
summary(lm(log(William) ~ O3 * SO2 * Stress, data = case1402))
```

```
##
## Call:
## lm(formula = log(William) ~ O3 * SO2 * Stress, data = case1402)
##
## Residuals:
##       Min       1Q     Median        3Q       Max
## -0.222274 -0.067130  0.007202  0.054343  0.213667
##
## Coefficients:
##                             Estimate Std. Error t value
## (Intercept)                   8.5184     0.1111  76.703
## O3                           -5.4844     1.5963  -3.436
## SO2                          -1.0567     3.0414  -0.347
## StressWell-watered            0.3605     0.1571   2.295
## O3:SO2                      -36.8530    43.7175  -0.843
## O3:StressWell-watered        -2.5543     2.2576  -1.131
## SO2:StressWell-watered       -3.6102     4.3012  -0.839
## O3:SO2:StressWell-watered    45.2629    61.8259   0.732
##                             Pr(>|t|)
## (Intercept)                  < 2e-16
## O3                           0.00236
## SO2                          0.73158
## StressWell-watered           0.03162
## O3:SO2                       0.40831
## O3:StressWell-watered        0.27005
## SO2:StressWell-watered       0.41030
## O3:SO2:StressWell-watered    0.47183
##
## Residual standard error: 0.1152 on 22 degrees of freedom
## Multiple R-squared:  0.8507, Adjusted R-squared:  0.8032
## F-statistic: 17.91 on 7 and 22 DF,  p-value: 9.896e-08
```

Once again, the place to start is at the bottom of the table, where we see there is no indication of a three-way interaction. If we look at the rest of the table it suggests no two-way interactions either, just main effects of ozone and water stress but not of sulphur dioxide. However, as already shown, we need to remember that the estimate of each effect (row) in this table is made after controlling for all of the other effects (rows). How does this compare to the sequential ANOVA table?

```
anova(lm(log(William) ~ O3 * SO2 * Stress, data = case1402))
```

```
## Analysis of Variance Table
##
## Response: log(William)
##                 Df  Sum Sq Mean Sq F value     Pr(>F)
## O3               1 1.13812 1.13812 85.7564 4.793e-09
## SO2              1 0.26558 0.26558 20.0114 0.0001898
## Stress           1 0.23764 0.23764 17.9062 0.0003427
## O3:SO2           1 0.00281 0.00281  0.2116 0.6499936
## O3:Stress        1 0.01013 0.01013  0.7632 0.3917698
## SO2:Stress       1 0.00238 0.00238  0.1790 0.6763797
## O3:SO2:Stress    1 0.00711 0.00711  0.5360 0.4718342
## Residuals       22 0.29197 0.01327
```

The ANOVA table agrees that the interactions are all unimportant (again, to be thorough, we might want to use drop1() to omit the three-way interaction and then reassess the two-way interactions, but if you try that you will see that it makes no difference in this case). However, the F-test of the main effect of sulphur dioxide now gives a stronger indication that it has an effect alongside those of ozone and water stress. We will take a closer look at the mismatch between the table of coefficients and the ANOVA table for the model with interactions in Chapter 14, but in this case model simplification to omit clearly unimportant higher-order interactions produces an ANOVA table for the minimal adequate model that is in line with the table of coefficients:

```
display(lm(log(William) ~ O3 + SO2 + Stress, data = case1402))
```

```
## lm(formula = log(William) ~ O3 + SO2 + Stress, data = case1402)
##                      coef.est coef.se
## (Intercept)            8.63     0.06
## O3                    -7.14     0.74
## SO2                   -3.77     0.80
## StressWell-watered     0.18     0.04
## ---
## n = 30, k = 4
## residual sd = 0.11, R-Squared = 0.84
```

```
anova(lm(log(William) ~ O3 + SO2 + Stress, data = case1402))
```

```
## Analysis of Variance Table
##
## Response: log(William)
##             Df  Sum Sq Mean Sq F value     Pr(>F)
## O3           1 1.13812 1.13812  94.119 3.969e-10
## SO2          1 0.26558 0.26558  21.963 7.690e-05
## Stress       1 0.23764 0.23764  19.652 0.0001501
## Residuals   26 0.31440 0.01209
```

In summary, the individual ANCOVAs and the three-way general linear model all support clear negative effects of water stress, ozone, and sulphur dioxide on soya bean yields but give no indication of interactive effects (as an exercise, you can compare the response of the Forrest variety).

13.6 Summary

In general, we have seen how ANCOVA of designed experiments combines one categorical variable and one continuous explanatory variable. Thanks to the concept of general linear models, this mix of regression and ANOVA can be performed with the lm() function just like the simpler individual analyses. Unfortunately, as we'll explore in the next chapter, the interpretation of the ANOVA and summary tables gets trickier and the equivalence of the F- and t-tests in these tables sometimes breaks down since the same comparisons are not being performed (despite what the row labels in the table sometimes imply). This means that we will usually have to fit a series of nested models and compare the results. When the results agree, it doesn't really matter if we present a more complex or a simpler version (and we can present whichever is more suited to the goals of any particular analysis), but when they differ in their details we will need to present multiple models and explain how and why they differ. As ever, for us the statistics is a means to an end and it is the biological interpretation that should remain as the main focus.

13.7 References

Anderson, D.R. (2008) *Model Based Inference in the Life Sciences*. Springer.
Ramsey, R.R. & Schafer, D. (2013) *The Statistical Sleuth*. Brooks/Cole Cengage Learning.

Linear Model Complexities

14.1 Introduction

In introducing linear-model analysis, I have tried to keep things as straightforward as possible. We began with the simplest types of linear model: one-way ANOVA and its simple linear regression equivalent. However, once we encountered more complex ANOVA and ANCOVA designs some complexities arose that we then skipped over. This chapter explores these complexities of linear-model analysis and some additional ones that arise with unbalanced designs—those with unequal numbers of replicates in the different treatment groups.

14.1.1 R PACKAGES

```
library(arm)
library(ggplot2)
library(Sleuth3)
```

The New Statistics with R: An Introduction for Biologists. Second Edition. Andy Hector,
Oxford University Press. © Andy Hector 2021. DOI: 10.1093/oso/9780198798170.003.0014

14.2 Analysis of variance for balanced designs

To explore the analysis of balanced designs, let's return to the example
of factorial ANOVA from earlier. In this chapter I am going to call the
dataframe 'Balanced'—here are the first few rows:

```
Balanced <- read.table("Data_Factorial.txt", header = TRUE)
head(Balanced)
```

```
##    fert_ light_  fl_ biomass
## 1    F-     L- F-L-   254.2
## 2    F-     L- F-L-   202.0
## 3    F-     L- F-L-   392.4
## 4    F-     L- F-L-   455.3
## 5    F-     L- F-L-   359.1
## 6    F-     L- F-L-   386.5
```

First, let's fit the model from the earlier chapter again and take a look at
the ANOVA table output:

```
anova(lm(biomass ~ light_ + fert_ + light_:fert_, data = Balanced))
```

```
## Analysis of Variance Table
##
## Response: biomass
##                Df Sum Sq Mean Sq F value    Pr(>F)
## light_          1  96915   96915 11.3131  0.001345
## fert_           1 319889  319889 37.3413 8.019e-08
## light_:fert_    1  36409   36409  4.2501  0.043587
## Residuals      60 513998    8567
```

The model formula asks about the effect of light, the effect of fertilizer,
and the interactive effect of the two. One important thing to know about
classical ANOVA tables is that the different effects are assessed *sequentially*,
with the order determined by the model formula. Of course, we could

rewrite this model reversing the order of light and fertilizer treatments—does altering the sequence of the explanatory variables in the linear-model formula make a difference to the ANOVA table?

```
anova(lm(biomass ~ fert_ + light_ + light_:fert_, data = Balanced))
```

```
## Analysis of Variance Table
##
## Response: biomass
##              Df Sum Sq Mean Sq F value    Pr(>F)
## fert_         1 319889  319889 37.3413 8.019e-08
## light_        1  96915   96915 11.3131  0.001345
## fert_:light_  1  36409   36409  4.2501  0.043587
## Residuals    60 513998    8567
```

No—with a balanced design with equal numbers of replicates for the treatment combinations, the values of the sums of squares are unaffected by the order of terms in the model formula. The balanced design ensures the property of *orthogonality*, which means that the two (or more) explanatory variables can be independently assessed—their effects do not depend on each another (Box 14.1).

Box 14.1 - Balance and orthogonality

An experimental design or analysis is said to be orthogonal when the effects of the response variables are uncorrelated. This makes the sums of squares calculated for the two variables independent of the order in which they are included in the model formula. Balanced experimental designs with equal numbers of replicates for each treatment combination ensure orthogonality. Orthogonality is lost when replication is unequal across the different treatment combinations—an extreme form is where a treatment combination is absent, producing an empty cell in the matrix of the experimental design. There is one special case when designs can be unbalanced but orthogonal—this is possible when replication is unequal but in a constant proportion (Grafen and Hails 2002). For example, imagine a design with two treatments and two blocks where the numbers of replicates of the treatments per block are unequal but in the constant proportion of 1 versus 2 in the first block but 2 versus 4 in the second.

The independence of the values of the sums of squares (and everything that follows from them) makes things relatively simple, and it is why balanced, orthogonal designs are so strongly recommended. However, it is not always possible to design balanced studies, or studies can become unbalanced due to the unintentional loss of replicates. Comparisons in the analysis of observational data are unlikely to be orthogonal—sadly, nature rarely provides us with balanced designs. Unfortunately, the analysis of non-orthogonal designs is more complex. To explore this we can revisit a hypothetical example from an article on the analysis of unbalanced designs.

14.3 Analysis of variance with unbalanced designs

The example comes from a paper by Shaw & Mitchell-Olds (1993). The made-up data set comprises the final height of plants (the imaginary study organisms) as the response variable, the experimental removal (or not) of neighbours as a first explanatory variable, and the initial size of the target plants as a second. Both explanatory variables are factors, each with two levels—initial sizes are recorded only as 'small' or 'large'. The design is therefore a fully factorial design comprising two factors—each with two levels—crossed so that all four possible combinations (or 'cells' in a tabular representation of the design) are present. The design is unbalanced and non-orthogonal because the different combinations have different numbers of replicates, but no cells are empty. Since the (made-up) data set has only 11 values, we can easily create it as follows:

```
height <-
  c(50, 57, 91, 94, 102, 110, 57, 71, 85, 105, 120)
size <-
  c(rep("Small", 2), rep("Large", 4), rep("Small", 3),
    rep("Large", 2))
treatment <- c(rep("Control", 6), rep("Removal", 5))
unbalanced <- data.frame(height, size, treatment)
unbalanced
```

```
##    height  size treatment
## 1      50 Small   Control
## 2      57 Small   Control
```

```
## 3       91 Large    Control
## 4       94 Large    Control
## 5      102 Large    Control
## 6      110 Large    Control
## 7       57 Small    Removal
## 8       71 Small    Removal
## 9       85 Small    Removal
## 10     105 Large    Removal
## 11     120 Large    Removal
```

The two-factor design can be summarized as a table with four cells—there are no empty cells, but the design is unbalanced with differing numbers of replicates for the four combinations of the two explanatory variables:

```
table(size, treatment)
```

```
##           treatment
## size      Control Removal
##    Large        4       2
##    Small        2       3
```

A model-fitting treatment first produces the following ANOVA table:

```
mod_TxS <- lm(height ~ treatment * size, data = unbalanced)
anova(mod_TxS)
```

```
## Analysis of Variance Table
##
## Response: height
##                Df Sum Sq Mean Sq F value    Pr(>F)
## treatment       1   35.3    35.3  0.3309 0.5831478
## size            1 4846.0  4846.0 45.3658 0.0002686
## treatment:size  1   11.4    11.4  0.1068 0.7533784
## Residuals       7  747.8   106.8
```

Reversing the order to put size first produces this alternative:

```
mod_SxT <- lm(height ~ size * treatment, data = unbalanced)
anova(mod_SxT)
```

```
## Analysis of Variance Table
##
## Response: height
##                 Df Sum Sq Mean Sq F value    Pr(>F)
## size             1 4291.2  4291.2 40.1718 0.0003896
## treatment        1  590.2   590.2  5.5249 0.0510495
## size:treatment   1   11.4    11.4  0.1068 0.7533784
## Residuals        7  747.7   106.8
```

First the good news—the value of the sum of squares for the residual is the same in both tables. This is also true for the interaction (note that we only have a single two-way interaction here; with more explanatory variables we might have higher-order interactions too). However, because sums of squares are calculated sequentially, with this unbalanced data set the order of the explanatory variables in the linear-model formula affects the values for the two main effects. This is also true if we remove the interaction from the model and assess the effect of treatment before size:

```
mod_TS <- lm(height ~ treatment + size, data = unbalanced)
anova(mod_TS)
```

```
## Analysis of Variance Table
##
## Response: height
##             Df Sum Sq Mean Sq F value    Pr(>F)
## treatment    1   35.3    35.3  0.3725    0.5586
## size         1 4846.0  4846.0 51.0676 9.746e-05
## Residuals    8  759.2    94.9
```

Or, if we reverse the order to consider size before treatment:

```
mod_ST <- lm(height ~ size + treatment, data = unbalanced)
anova(mod_ST)
```

```
## Analysis of Variance Table
##
## Response: height
##              Df Sum Sq Mean Sq F value    Pr(>F)
## size          1 4291.2  4291.2 45.2208 0.0001489
## treatment     1  590.2   590.2  6.2193 0.0372980
## Residuals     8  759.2    94.9
```

In both models initial size is clearly related to final height, as indicated by the high F-value. But initial size was not the explanatory variable of interest. Rather, it is a 'covariate' added to the model to control for differences in initial size (if the initial size had been the same across treatments, we wouldn't need to include it in the model). The neighbour removal treatment was the explanatory variable the experiment was designed to address. Unfortunately, the lack of balance has a substantial effect on the values of the sums of squares for treatment in the two alternative models. When entered into the model formula first, treatment has a low F-value and a high P-value. However, when it is put after initial size the F-value is much higher (and the P-value smaller). We now have a dilemma: the two models produce different results—what do we do? Well, a better experiment (more balanced) for one thing! To be fair, this is a deliberately small hypothetical example made up to illustrate the issues of unbalanced ANOVA, and these types of issue are very common in observational data sets where there is no control over the design. The first point is that if we conduct both analyses, we should be honest that the results of this analysis depend on the order of the treatments in the model (we should not pick one and forget about the other!). Some statisticians have tried to get around this ambiguity by devising different types of 'adjusted' sums of squares that produce order-independent values even for unbalanced designs, but the balance of opinion seems to generally recommend against this solution (Box 14.2). However, in this case I would argue that one of these models fits the intentions of this analysis better than the other: the model that controls for size and then assesses the effect of treatment seems to make more sense. The whole point of including the covariate for initial size was to adjust for its effects before assessing the effect of treatment. In other words, with a bit of careful consideration beforehand, I think it would have been possible to select the sequential model that controls for size before assessing treatment effects as the most appropriate.

Box 14.2 - Type III sums of squares

The order dependence of the sums of squares in the analysis of unbalanced designs is a frustrating complexity. In response, various ways of drawing up composite tables of different types of 'adjusted' sums of squares have been devised (Hector et al. 2010). In particular, type III sums of squares were invented to provide values that are not order-dependent *even with unbalanced data sets*. Type III sums of squares (and their even more complex relatives) were a highly debated topic for a long time. The dispute seems to have largely burned itself out now, with most statisticians generally avoiding type III sums of squares (although they still have their backers). Notably, they are not available in the base distribution of R (some packages provide them, although usually as an option rather than the default). Many of the statisticians cited in this book recommend against them (including John Nelder and Bill Venables), with Maindonald and Braun concluding that 'Most who know what they are doing avoid Type 3 tests.

14.4 ANOVA tables versus coefficients: When *F* and *t* can disagree

Above, we have demonstrated the order dependence of ANOVA tables for unbalanced, non-orthogonal designs and therefore the F-tests that they contain. Is the same true for the estimates and t-tests in the table of coefficients? Here are the estimates and t-tests for the model (without the interaction term) that gives priority to treatment:

```
summary(mod_TS)
```

```
##
## Call:
## lm(formula = height ~ treatment + size, data = unbalanced)
```

```
##
## Residuals:
##      Min       1Q   Median      3Q      Max
## -13.1053  -6.2105   0.8947   4.7895  14.8947
##
## Coefficients:
##                     Estimate Std. Error t value Pr(>|t|)
## (Intercept)            98.58       4.47  22.055 1.89e-08
## treatmentRemoval       15.26       6.12   2.494   0.0373
## sizeSmall             -43.74       6.12  -7.146 9.75e-05
##
## Residual standard error: 9.741 on 8 degrees of freedom
## Multiple R-squared:  0.8654, Adjusted R-squared:  0.8318
## F-statistic: 25.72 on 2 and 8 DF,  p-value: 0.0003281
```

For comparison, here is the summary for the model that gives priority to initial size:

```
summary(mod_ST)
```

```
##
## Call:
## lm(formula = height ~ size + treatment, data = unbalanced)
##
## Residuals:
##      Min       1Q   Median      3Q      Max
## -13.1053  -6.2105   0.8947   4.7895  14.8947
##
## Coefficients:
##                     Estimate Std. Error t value Pr(>|t|)
## (Intercept)            98.58       4.47  22.055 1.89e-08
## sizeSmall             -43.74       6.12  -7.146 9.75e-05
## treatmentRemoval       15.26       6.12   2.494   0.0373
##
## Residual standard error: 9.741 on 8 degrees of freedom
## Multiple R-squared:  0.8654, Adjusted R-squared:  0.8318
## F-statistic: 25.72 on 2 and 8 DF,  p-value: 0.0003281
```

In contrast to the sums of squares in the ANOVA tables, the estimates of the coefficients are unaffected by the order of the explanatory variables in the linear-model formula. This is because they are not calculated sequentially; instead, the estimate for each term is made after controlling for the effects of all other variables in the model (including any interactions). For this no-interaction model with just the two main effects, the row in the summary table for treatment corresponds to the row in the ANOVA table where treatment is last in the model formula, and the same is true for size—as you can see from the match between the P-value for the t-test with the P-value or the F-test *when each variable is in last place in the sequential ANOVA*.

14.5 Marginality of main effects and interactions

The two linear-model analyses with interactions explored above have different outcomes—in one case the analysis supported the interaction and in the other example it did not. This touches on the difficult issue of whether it is better to retain all terms in a statistical analysis or whether we should try to simplify models by removing terms that appear unimportant. This includes the issue of whether to remove unsupported interactions or whether to retain them regardless. You will read different advice in different places. When I was first taught statistics (by Mick Crawley at Imperial College), common advice was to simplify to a 'minimal adequate model' that included only terms that had clearly detectable effects (plus 'marginal' terms when you want to be inclusive). However, more recent sources— Gelman & Hill (2006) for example—suggest keeping unimportant interactions in the model regardless, so long as they do not behave unexpectedly (have an unexpected sign, for example)—their book gives some examples. We have already encountered this issue in Chapter 13, where we extended the typical ANCOVA model with one continuous variable and one factor into a general linear model that added a third variable (a factor in this case). One complexity we skipped over was exactly the issue raised here—that the importance of the main effects varied depending on whether interactions

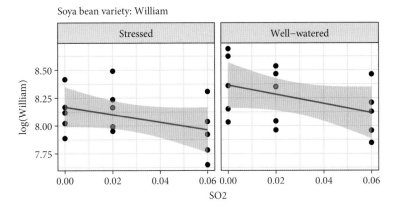

Figure 14.1 Yield (natural-log transformed) of William-variety soya beans as a function of sulphur dioxide and water stress.

were present in the model or not. Let's finish this chapter by revisiting that example. As it happens, this complexity isn't a product of having three variables in the model; it exists even with a model with just the two variables for the effects of water stress and sulphur dioxide (Fig. 14.1):

```
Fig14_1 <-
  qplot(data = case1402, x = SO2, y = log(William)) +
  facet_wrap(. ~ Stress) +
  geom_smooth(method = "lm", formula = y ~ x) +
  ggtitle("Soya bean variety: William")
Fig14_1
```

Let's start with the table of coefficients (which are unaffected by the order of terms in the model formula, remember):

```
summary(lm(log(William) ~ SO2 * Stress, data = case1402))
```

```
##
## Call:
## lm(formula = log(William) ~ SO2 * Stress, data = case1402)
##
## Residuals:
##        Min        1Q     Median        3Q        Max
```

```
## -0.33175 -0.17513 -0.00192   0.17254   0.39154
##
## Coefficients:
##                          Estimate Std. Error t value
## (Intercept)               8.16740    0.08926  91.499
## SO2                      -3.41525    2.44456  -1.397
## StressWell-watered        0.19703    0.12624   1.561
## SO2:StressWell-watered   -0.71339    3.45713  -0.206
##                          Pr(>|t|)
## (Intercept)               <2e-16
## SO2                        0.174
## StressWell-watered         0.131
## SO2:StressWell-watered     0.838
##
## Residual standard error: 0.2362 on 26 degrees of freedom
## Multiple R-squared:  0.2585, Adjusted R-squared:  0.173
## F-statistic: 3.022 on 3 and 26 DF,  p-value: 0.04769
```

Starting at the bottom, we see that there is no indication of an interaction. If we assess the coefficients for the main effects of SO_2 and water stress in the presence of this interaction, they don't provide any clear support for effects of these variables either. So, if we looked only at this table, the impression we would get is of no effects of water or sulphur dioxide—either alone or in combination. However, if we look at an ANOVA table summary for this analysis then we get a different picture:

```
anova(lm(log(William) ~ SO2 * Stress, data = case1402))
```

```
## Analysis of Variance Table
##
## Response: log(William)
##              Df  Sum Sq  Mean Sq F value  Pr(>F)
## SO2           1 0.26558 0.265581  4.7617 0.03833
## Stress        1 0.23764 0.237642  4.2608 0.04911
## SO2:Stress    1 0.00238 0.002375  0.0426 0.83812
## Residuals    26 1.45014 0.055775
```

The ANOVA table confirms the lack of any indication of an interaction, but the F-tests for the main effects of water stress and sulphur dioxide now support main effects of these variables (according to the $P < 0.05$ convention), even in the presence of the non-significant interaction. Earlier we saw that for the one-way ANOVA of Darwin's maize data the ANOVA table F-tests and summary table t-tests produced identical P-values, but for this ANOVA things are different: while the P-value for the interaction is the same in both tables, the tests for the upper two rows disagree. Note that we are not making the mistake of interpreting main effects in the presence of a significant interaction effect here—the interaction is nowhere near the conventional significant level. What is going on?

Unfortunately, there are various situations where the ANOVA table F-tests and summary table t-tests can differ. As we noted above, equivalency requires orthogonality (Box 14.1). However, non-orthogonality is not the explanation for the mismatch here, since in the ANCOVA with sulphur dioxide there are balanced numbers of replicates of the pollutant treatment for the two levels of water stress and the values for the sums of squares in the ANOVA table are unaffected by the position of the variables in the model formula (try reversing the order of the terms in the model formula and re-fitting the models to see this for yourself).

In this case the mismatch comes about because the tests in the summary table are not exactly the same as the tests performed in the ANOVA table, despite the similar or identical row names. The ANOVA table has a first row for the main effect of SO_2, a second row for the main effect of water stress, and a third row for the interaction (plus the ever-present residual error). In contrast, the table of coefficients has four rows. The first row is the intercept of the regression with SO_2 in the water-stressed treatment level, the second row is the slope of this regression, and the third and fourth rows give the *differences* in the intercept and slope of the regression in the well-watered treatment. So, while both tables have a row labelled 'SO2', in the ANOVA table this row quantifies how much of the variance is explained by the sulphur dioxide gradient and the F-value gives the ratio of this variance relative to the unexplained noise (as quantified by the residual mean square). However, in the summary table this row estimates the slope of the regression of yield on sulphur dioxide concentration for the well-watered treatment, after adjusting for the effects of the other terms in the model, including the (non-significant) interaction term. In short, when analyses are simple, the summary and ANOVA tables will often perform equivalent tests and the resulting *P*-values will be the same. However, as analyses become more complex due to the presence of interactions or a loss of orthogonality (or both), the rows in the two tables will often not perform equivalent tests (despite sometimes having the same row label) and there will be a difference in the results. Sometimes the qualitative outcome will be the same, but sometimes not (this is a problem with using arbitrary cut-off points, where examples close to the boundary can be nudged one way or the other by changes in the model such as the inclusion or omission of an interaction term). This means that we have to have a good understanding of the models we are fitting, explore alternative models, and explain any ambiguities in the results.

For example, one thing to notice here is that the interaction term explains less variation than we would expect, even if there was no interaction and the interaction sum of squares was pure noise. We can see this from the small values for the sums of squares, which result in a variance that is much smaller than the error mean square and an *F*-value smaller than expected (we would expect a value of one if there were absolutely no interaction and we could estimate the noise perfectly), i.e. a negative variance component. Negative variance components have two interpretations. On the one hand, a negative variance component could indicate that something is wrong with the data and the analysis is not working as expected. On the other hand, because we cannot quantify the signal and noise exactly—only estimate them based on samples—the small variance could be an underestimate due to sampling variation. Sometimes a negative variance component will lead to the discovery of a problem with a data set, but with the huge number of analyses being performed around the world there will be lots of sampling error, and usually we have to put them down to this cause. We did not collect these data on soya bean yields, but the design and analysis look good, so this is most likely one of those cases where the negative variance component has resulted due to sampling variation. Happily, if we remove the interaction term then the ANOVA table and the table of coefficients are in agreement, with both providing support for effects of sulphur dioxide and water that seem biologically sensible: reducing water stress increases yield, while the pollutant has a negative effect:

```
summary(lm(log(William) ~ SO2 + Stress, data = case1402))
```

```
##
## Call:
## lm(formula = log(William) ~ SO2 + Stress, data = case1402)
##
## Residuals:
##      Min      1Q   Median      3Q      Max
## -0.3222  -0.1711  -0.0051  0.1737   0.3892
##
## Coefficients:
##                     Estimate Std. Error t value Pr(>|t|)
## (Intercept)          8.17692    0.07507 108.920   <2e-16
## SO2                 -3.77195    1.69764  -2.222   0.0349
## StressWell-watered   0.17800    0.08469   2.102   0.0450
##
## Residual standard error: 0.2319 on 27 degrees of freedom
## Multiple R-squared:  0.2573, Adjusted R-squared:  0.2023
## F-statistic: 4.677 on 2 and 27 DF,  p-value: 0.01803
```

```
anova(lm(log(William) ~ SO2 + Stress, data = case1402))
```

```
## Analysis of Variance Table
##
## Response: log(William)
##           Df  Sum Sq  Mean Sq F value  Pr(>F)
## SO2        1 0.26558 0.265581  4.9367 0.03486
## Stress     1 0.23764 0.237642  4.4174 0.04504
## Residuals 27 1.45252 0.053797
```

This example suggests that if we retain unimportant interactions in the model we run the risk of missing some of the main effects. It suggests that to be safe, we need to assess main effects with and without the interaction in the model to be sure to get the full picture. One point to note here is the relatively small sample size in this example ($n = 30$) compared with the examples in Gelman & Hill (2006), which have much larger sample sizes and therefore greater power to detect main effects even with unimportant interactions left in the model. Whether or not to retain non-significant terms in a model can also depend on the aim. When the aim is prediction, it can make sense to be cautious and retain non-significant terms (so long as they are not behaving strangely—with a sign counter to expectations, for example). However, when the focus is on hypothesis testing and discovering the 'true' model, simplifying to a minimal adequate model may be more consistent with the original aims.

14.6 Summary

Statisticians strongly recommend balanced, orthogonal designs because their analysis lacks several complexities and ambiguities (marginality and order dependence) that arise when these desirable properties are absent (as in most observational data sets). When these complexities are present, some careful forethought can save a lot of work and help to lay out an efficient route through the analysis that is consistent with the original aims and minimizes ambiguity. This includes whether the focus is on prediction or on hypothesis testing, explanation, and understanding.

14.7 References

Gelman, A. & Hill, J. (2006) *Data Analysis Using Regression and Multilevel/Hierarchical Models*. Cambridge University Press.

Grafen, A. & Hails, R. (2002) *Modern Statistics for the Life Sciences*. Oxford University Press.

Hector, A., von Felten, S., & Schmid, B. (2010) Analysis of variance with unbalanced data: An update for ecology & evolution. *Journal of Animal Ecology* 79: 308–16.

Shaw, R.G. & Mitchell-Olds, T. (1993) ANOVA for unbalanced data: An overview. *Ecology* 74: 1638–45.

Generalized Linear Models

15.1 GLMs

In Chapter 7 we used normal least squares to perform linear regression
analysis, implemented with the lm() function. However, the simple
linear regression analysis could not accommodate curvature in the relationship and the normal least squares assumption of constant variance was
infringed. A more flexible approach would model the mean and variance
separately, and this is exactly what GLMs do. The aim of this chapter is to
use GLMs and the maximum likelihood methods on which they are based
to extend the linear regression example further. The analysis was developed
by the statistician Bill (William) Venables with input from John Nelder, one
of the originators of GLMs.

15.1.1 R PACKAGES

```
library(cowplot)
library(ggfortify)
library(ggplot2)
library(MASS)
library(SemiPar)
```

The New Statistics with R: An Introduction for Biologists. Second Edition. Andy Hector,
Oxford University Press. © Andy Hector 2021. DOI: 10.1093/oso/9780198798170.003.0015

15.2 The trouble with transformations

In the exercise on linear regression, we began with the following model:

```
library(SemiPar)
data(janka)
janka.ls1 <- lm(hardness ~ dens, data = janka)
```

However, as we discovered, analysis of the Janka data in its raw form led to infringements of some of the model assumptions. The two key issues are that there is some curvature in the relationship that is not captured, and that the variance is not constant—it increases with the mean (refer back to the linear regression diagnostic plots). One way to deal with these issues is to transform the response variable. For example, as we have an upward bend in our positive relationship one possibility would be to apply the commonly used square-root transformation to the timber hardness data to reduce the high values more than the small ones in an attempt to get them to fall closer to a straight-line relationship with density. We could create a new column of transformed data and add it to the dataframe, or we can perform the transformation directly within the linear-model function:

```
janka.sqrt <- lm(sqrt(hardness) ~ dens, data = janka)
```

We can add the line from this regression to a graph of the transformed data (Fig. 15.1):

```
fig15_1 <- qplot(data = janka, x = dens, y = sqrt(hardness)) +
  geom_smooth(method = "lm", formula = y ~ x)
fig15_1
```

Comparing the residuals with those from the linear regression analysis of the untransformed data shows that we have improved the linearity but that the residuals still increase with the mean (use the plot() function if you do not have the ggfortify package) (Fig. 15.2):

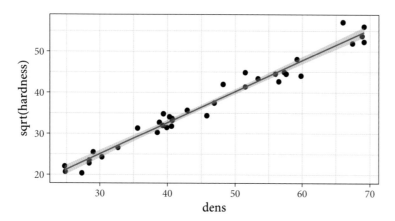

Figure 15.1 Linear regression of the square-root-transformed timber hardness data.

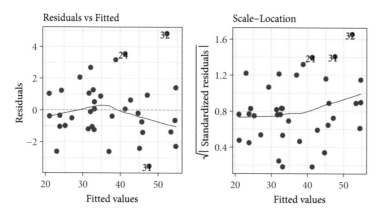

Figure 15.2 Diagnostic plots of the residuals of the square-root-transformed Janka timber hardness data.

```
fig15_2 <- autoplot(janka.sqrt, which = c(1, 3), ncol = 2)
fig15_2
```

Another common transformation that we could try is to log the data (using the natural log in this case):

```
janka.log <- lm(log(hardness) ~ dens, data = janka)
```

Now we can see that while the variance is more constant, the transformation has introduced curvature in the opposite direction—it has overcorrected, if you like (the red dashed smoother helps pick out the curvature)) (Fig. 15.3):

```
fig15_3 <-
  qplot(data = janka, x = dens, y = log(hardness)) +
  geom_smooth(method = "lm", formula = y ~ x) +
  geom_smooth(se = FALSE, colour = "red", linetype = "dashed")
fig15_3
```

The residual plots give a more detailed view that confirms that while the variance is more constant, the transformation has introduced curvature—this transformation has fixed one problem but introduced another (Fig. 15.4):

```
fig15_4 <- autoplot(janka.log, which = c(1, 3), ncol = 2)
fig15_4
```

Of course, we could extend this approach to analysing the log-transformed data using a polynomial regression that supplements the linear term for density with a quadratic term (density squared) in the model formula (note that raising the values of density to the power 2 has to be done together with the I() function):

```
janka.quad <- lm(log(hardness) ~ dens + I(dens^2), data = janka)
```

With polynomial regression, each additional term (quadratic, cubic, etc.) adds a turning point to the relationship—in this case adding the quadratic term turns the straight line into a simple curve (Fig 15.5):

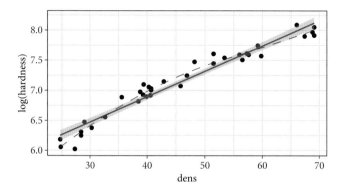

Figure 15.3 Linear regression of the log-transformed timber hardness data.

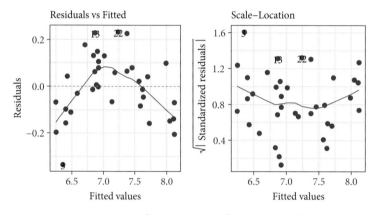

Figure 15.4 Diagnostic plots of the residuals of the log-transformed Janka timber hardness data.

```
fig15_5 <- qplot(data = janka, x = dens, y = log(hardness)) +
  geom_smooth(method = "lm", formula = y ~ x + I(x^2))
fig15_5
```

However, while this regression addresses both the curvature and the increasing variance (check the diagnostic plots yourself), we now have a more complex model. Furthermore, polynomial regressions come with the health warning that they often extrapolate poorly (and sometimes very badly in complex cases) beyond the range of the data.

In summary, a square-root transformation of the data produced a linear relationship but the variance still increased with the mean. A log

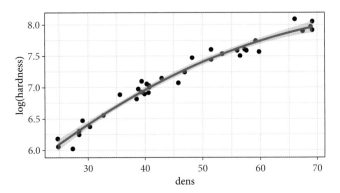

Figure 15.5 Polynomial regression of the log-transformed timber hardness data.

transformation of the data produced constant variance but 'overcorrected' and introduced curvature in the opposite direction. Neither transformation led to a regression that adequately modelled both the mean and the variance. GLMs take a more flexible approach that models the mean and the variance separately.

15.3 The Box–Cox power transform

In this example we have decided to try the square-root and natural-log transformations, but perhaps there are better alternatives? The MASS package, by Venables and Ripley, has a handy function for applying the Box–Cox family of transformations (Box 15.1).

Box 15.1 - The Box–Cox family of transformations

This approach gets its distinctive name from its inventors, George Box and David Cox. The approach is implemented in R with the boxcox() function, which is part of the MASS package that complements *Modern Applied Statistics with S* by Venables & Ripley (2002). The transformations are done by raising the data to a power (lambda) that is systematically varied. For example, when lambda = 2 the data are squared, when lambda = 0.5 they are square-rooted, and so on. The data raised to the power one are untransformed. Since any number raised to the power of zero equals one, a special behaviour is defined that integrates smoothly with the other transformations:

lambda = 0 is the natural-log transformation. The R output for the boxcox() function plots the maximum likelihood surface (the curve) together with a maximum-likelihood-based 95% confidence interval. The interval is produced by dropping down 1.92 log-likelihood units from the maximum likelihood value, moving horizontally left and right (the dotted line labelled 95% in Fig. 15.6) until the likelihood surface is met and then dropping down from these points as shown by the outer vertical dotted lines in the same figure. The value of 1.92 derives from the critical value of chi-squared (the deviance, the generalized equivalent of sums of squares, is distributed approximately following chi-squared) with 1 DF and at the 95% level of confidence, divided by 2 (for a two-tailed test).

Since the Box–Cox transformation uses maximum likelihood, it provides a convenient way to introduce that approach too (Box 15.2).

Box 15.2 - Maximum likelihood

The idea is that, for a given data set and with a specified statistical model, maximum likelihood estimation finds the values of the parameters of the statistical model that are most likely to reproduce the observed data: in other words, the parameter values that give the best fit of the model to the data. Hopefully, this will sound familiar from the way least squares finds the line of best fit by minimizing the sum of the squared differences. Indeed, for data with normally distributed errors, the particular form of maximum likelihood analysis is none other than the normal least squares we have already met. However, the normal distribution will not be a good model for some forms of data: binary data are an obvious example that we will look at later. Maximum likelihood provides a more general approach that applies to a broader range of data than the more restrictive normal least squares—hence *generalized* linear models. The more general method of calculation that underlies GLMs is known as iterative weighted least squares (IWLS). The details of that are beyond the scope of this book, but we can think of the IWLS versions for non-normal data as analogous to the normal least squares process we are familiar with. The details of the calculations for different types of data (the different likelihood functions) are in standard statistics texts, while Mick Crawley's *The R Book* (Crawley 2007) and Ben Bolker's *Integrating Ecological Models and Data in R* (Bolker 2008) give great introductions for non-statisticians. One strength of normal least squares is that the equations that underlie the method have exact solutions. In contrast, maximum likelihood methods are iterative and approximate: essentially, we try out many potential lines of best fit using algorithms that gradually home in on what looks to be the best solution. To avoid very small numbers, the calculations are done with the log-likelihood (calculations that are multiplicative on the likelihood scale become additive with log-likelihoods). The generalized version of the sum of squares is called the deviance. The deviance is defined in terms of comparing pairs of models where one is more complex and the other a simplified version (as done in R using the anova() function). The deviance is twice the difference in the log-likelihoods of the more complex and simplified models (expressed as a negative value). The deviance approximately follows the chi-squared distribution (with the number of degrees of freedom equal to the number of parameters of the model), which allows the calculation of levels of probability and confidence, just as the normal distribution does when using normal least squares.

The Box–Cox method tries out many transformations and assesses which one produces the best fit to the data (so it is only the 'best' transformation by this criterion). The best transformation corresponds to the value of lambda that produces the maximum value of the log-likelihood, as indicated in Fig. 15.6 together with a 95% CI:

```
library(MASS)
Fig15_6 <- boxcox(janka.ls1)
```

The results of the Box–Cox transformation suggest that a value of lambda of around 0.5 will give the best fit to the data, so we can proceed using the square-root transformation, or rather the GLM equivalent.

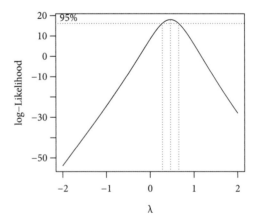

Figure 15.6 A likelihood surface for the Box–Cox function applied to the simple linear regression of the Janka timber hardness data.

15.4 Generalized Linear Models in R

GLMs were first proposed by Nelder and Wedderburn in the early 1970s, and *Generalized Linear Models* by McCulloch & Nelder (1983) is the standard reference. GLMs have three components:

1. A linear predictor.
2. A variance function.
3. A link function.

The first is the most familiar. In R, the linear predictor is what comes after the tilde (~) in our linear-model formula. In the Darwin's maize example, the linear predictor included 'pair' and pollination type. In the example in this chapter, it is wood density. The variance function models the variation in the data. This is also familiar to us, since normal least squares uses the normal distribution to model the residual variation. The difference here is that GLMs are not restricted to the normal distribution but make use of a wider range of distributions (most taken from the exponential family), including the Poisson, the binomial, and the gamma. The third component is the least familiar. The link function plays a role equivalent to the transformation in normal least squares models. However, rather than transforming the data, we transform the predictions made by the linear predictor. Commonly used link functions include the log, square root, and logistic functions.

First of all, let us compare the output when we fit the same model using the lm() and glm() functions. Our original normal least squares regression model produced the following ANOVA table output:

```
anova(janka.ls1)
```

```
## Analysis of Variance Table
##
## Response: hardness
```

```
##              Df   Sum Sq  Mean Sq F value     Pr(>F)
## dens          1 21345674 21345674  636.98 < 2.2e-16
## Residuals 34   1139366    33511
```

We can fit the same model using the glm function as follows:

```
janka.ml1 <- glm(hardness ~ dens, data = janka,
    family = gaussian(link = "identity"))
```

We could omit the family and link argument labels, but they make it clear that we are using the Gaussian (or normal) distribution to model the residual variability and the identity link to model the mean. The identity link is equivalent to performing no transformation, so no transformation is applied to the fitted values from the linear predictor. The GLM produces the following output from the anova() function:

```
anova(janka.ml1)
```

```
## Analysis of Deviance Table
##
## Model: gaussian, link: identity
##
## Response: hardness
##
## Terms added sequentially (first to last)
##
##
##        Df Deviance Resid. Df Resid. Dev
## NULL                     35    22485041
## dens    1 21345674        34     1139366
```

First, notice that although we are using the anova() function instead of the least squares ANOVA table produced when fitting models with the lm() function, we get the maximum likelihood analysis-of-deviance (ANODEV) table when using glm(). At first glance the outputs look more different than they actually are. First, remember that for GLMs using the normal distribution the deviance is none other than the sum of squares. The values of the SS for the density and the residuals appear as the deviance and residual deviance in the bottom row of the GLM output, alongside the degrees of freedom. The upper row of the ANODEV output contains the values of the total DF and total deviance found by summing the values given for the density and residuals (recall that the least squares ANOVA tables produced by R do not give the total). So, the values for the deviance here match the values from a conventional ANOVA table. However, there is no equivalent of the mean squares in the analysis of deviance. We will see in a moment how to get test statistics and *P*-values in the analysis-of-deviance table when we want them.

Now that we have demonstrated that the GLM function with the Gaussian variance function and identity link performs an analysis equivalent to the lm() function, we can see how to take fuller advantage of its greater flexibility. We need to model the mean and the variance. We know that in the normal least squares analysis the best fit was produced by the square-root transformation, as recommended by the Box–Cox results. However, with this transformation the variance increased as the mean increased. We can have the best of both worlds in our GLM by using a square-root link function in combination with a distribution in which the variance increases with the mean. It turns out that, for the gamma distribution, the variance increases as the square of the mean. We can fit a GLM with a square-root link and a gamma distribution variance function as follows:

```
janka.gamma <- glm(hardness ~ dens, data = janka,
    family = Gamma(link = "sqrt"))
```

We can then use qplot() to draw a graph of the resulting model (Fig. 15.7):

```
Fig15_7 <-
  qplot(data = janka, x = dens, y = hardness) +
  labs(x = "Density", y = "Hardness") +
  ggtitle("GLM, square-root link, Gamma variance") +
  geom_smooth(method = "glm", method.args = list(Gamma(link = "sqrt")))
Fig15_7
```

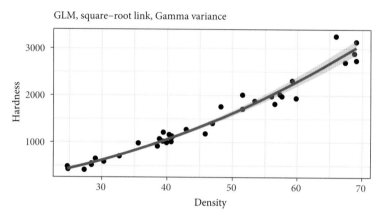

Figure 15.7 A GLM regression analysis using a square-root link function and modelling the increasing variance using the gamma distribution.

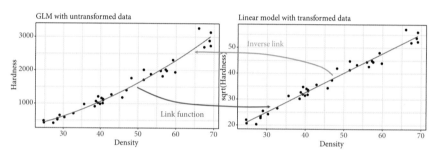

Figure 15.8 Visualization of the link function which relates the straight line from the linear model to the curve from the GLM.

Figure 15.8 might help to relate the GLM approach using a link function to the linear-model analysis of the square-root-transformed data.

To recap, the untransformed data (left-hand panel) show that the positive linear relationship has some upward curvature with increasing scatter. The linear regression of the square-root-transformed data (right-hand panel) linearizes the relationship by reducing the larger values more in absolute terms than the smaller values. In contrast, the GLM (left-hand panel) uses a link function to transform the linear relationship on the square-root scale to an upward-bending curve through the middle of the untransformed data. In the GLM analysis, the data are left alone and the transformation is applied to the predictions via the link function. The increasing variability is handled separately in the GLM by the gamma distribution variance function. You can think of the link function (and its opposite, the inverse link function) as mapping the straight line in the right-hand panel onto the curve in the left-hand panel.

We can get point estimates for the regression intercept and slope (remember the curve in the figures is linear on the scale of the link function, the square-root scale in this case) as follows:

```
coef(janka.gamma)
```

```
## (Intercept)          dens
##    1.8672884    0.7677963
```

Confidence intervals are extracted with the confint() function, but for GLMs they are calculated using maximum likelihood methods—likelihood profile intervals—that are more appropriate for GLMs than those from normal least squares methods (see the illustration of the Box–Cox transformation in Fig. 15.6). Note that since they are not based on the normal distribution, they are not constrained to be symmetric:

```
confint(janka.gamma)
```

```
## Waiting for profiling to be done...
```

```
##                    2.5 %     97.5 %
## (Intercept) 0.09706551 3.6699483
## dens        0.72361627 0.8122638
```

As we will explore further in the following chapters on GLMs, the same extractor functions are available for GLMs as for linear models fitted with the lm() function. Now that we have the gamma GLM, it can be used to calculate predictions as shown earlier for the linear regression.

15.5 Summary: Statistics

A simple normal least squares linear regression failed to capture the curvature in the relationship and infringed the assumption of approximately equal variability around the regression line: as is often the case, the variance increases with the mean. A square-root transformation of the Janka wood hardness data produces a linear relationship but the variability is still unequal. In contrast, log transformation equalizes the variability but generates its own curvature. Instead, we can use a more flexible GLM approach that can model the mean and variability independently. GLMs based on maximum likelihood use a link function to model the mean (in this case a square-root link) and a variance function to model the variability (in this case the gamma distribution, where the variance increases as the square of the mean).

15.6 Summary: R

- GLMs are fitted using the glm() function.
- The distribution is specified using the family argument.
- The link function is specified in parentheses after the name of the distribution.

15.7 References

Bolker, B. (2008) *Integrating Ecological Models and Data in R*. Princeton University Press.

Crawley, M.J. (2007) *The R Book*. Wiley.

McCulloch, P. & Nelder, J. (1983) *Generalized Linear Models*. Chapman & Hall.

Venables, W.N. & Ripley, B.D. (2002) *Modern Applied Statistics with S*. Springer.

GLMs for Count Data

16.1 Introduction

Count data are integers—whole numbers—for example, numbers of individuals, numbers of species, numbers of times an event occurred, etc. The starting point for a GLM analysis of count data is to use the Poisson distribution and a log link function. The log link function ensures that all predicted counts are positive by taking the exponential of the values generated by the linear predictor (antilogs are always positive values). For the Poisson distribution the variance is defined as equal to the mean, but, as always, this assumption needs to be examined since count data won't necessarily have this property in all cases. Count data where the Poisson distribution provides a good model usually have lots of zeros and small values. Just because data are counts, it does not follow that the Poisson distribution will necessarily provide a good model. Indeed, as the mean of the Poisson distribution increases, the distribution it converges towards the normal distribution.

The New Statistics with R: An Introduction for Biologists. Second Edition. Andy Hector, Oxford University Press. © Andy Hector 2021. DOI: 10.1093/oso/9780198798170.003.0016

16.1.1 R PACKAGES

```
library(arm)
library(ggplot2)
```

16.2 GLMs for count data

The example data are counts of grassland plant species in relation to levels of nitrogen deposition (kindly contributed by Carly Stevens). As we saw in Chapter 12, increasing nutrient inputs to grasslands usually results in a decline in their diversity. The factorial example was a controlled experiment looking at the change of diversity following nutrient enrichment—is the same true in surveys of grassland diversity in relation to the level of nitrogen pollution they receive? The data are in a file called Data_species_counts.txt:

```
Species <- read.table("Data_species_counts.txt", header = TRUE)
```

The data have just two columns, giving the level of nitrogen deposition (N_deposition) and the counts of the numbers of grassland plant species:

```
str(Species)
```

```
## 'data.frame':    74 obs. of  2 variables:
## $ N_deposition : num  8.56 7.7 8.28 8.14 10.99 ...
## $ Species_counts: int  20 17 25 18 20 10 13 14 15 15 ...
```

The N deposition data are continuous and the species counts are integers:

```
summary(Species)
```

```
##   N_deposition    Species_counts
## Min.   : 7.70    Min.   : 6.00
## 1st Qu.:14.26    1st Qu.:10.00
## Median :20.25    Median :13.00
```

```
##   Mean    :20.58    Mean    :13.91
##   3rd Qu.:27.11     3rd Qu.:15.00
##   Max.    :40.86    Max.    :27.00
```

A graph of the species count versus N deposition level shows a clear negative relationship, so we might be tempted to use a linear regression relationship as follows (Fig. 16.1):

```
Fig16_1 <-
   qplot(data = Species, x = N_deposition, y = Species_counts) +
   stat_smooth(method = "lm")
Fig16_1
```

At first glance, the linear regression appears to do a good job, but if extrapolated to higher nitrogen deposition levels it would predict negative species counts. The variability is also greater when species counts are higher, but the linear regression does not do a great job of modelling this, as we will see below. For comparison, we can fit a Poisson GLM as follows:

```
glm1 <- glm(Species_counts ~ N_deposition,
      family = poisson(link = "log"), data = Species)
```

Before we examine the model, we can try and assess whether a Poisson GLM is appropriate for these count data. The Poisson distribution is defined

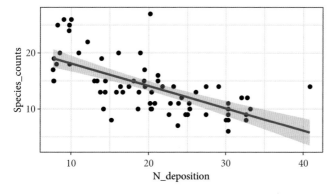

Figure 16.1 Relationship between the number of grassland plant species and nitrogen deposition modelled with a linear regression.

as having a variance equal to its mean. If this assumption is met, the ratio of residual deviance to residual DF (the dispersion) should be approximately 1:1 (although this ratio varies with the mean count to some degree, as shown in Venables & Ripley (2002)), which we can check from the bottom of the summary() output:

```
summary(glm1)
```

```
##
## Call:
## glm(formula = Species_counts ~ N_deposition,
##      family = poisson(link = "log"), data = Species)
##
## Deviance Residuals:
##      Min        1Q    Median        3Q       Max
## -2.1788   -0.6813   -0.1904    0.5825    3.1749
##
## Coefficients:
##                Estimate Std. Error z value Pr(>|z|)
## (Intercept)    3.210340   0.080655  39.803  < 2e-16
## N_deposition  -0.029436   0.003975  -7.405 1.31e-13
##
## (Dispersion parameter for poisson family taken to be 1)
##
##      Null deviance: 120.778  on 73  degrees of freedom
## Residual deviance:  64.707  on 72  degrees of freedom
## AIC: 396.2
##
## Number of Fisher Scoring iterations: 4
```

With a ratio of approximately 65:72, the residual deviance is as expected for the Poisson distribution, suggesting it makes an acceptable model for use with this data set. We can think of the Poisson GLM as equivalent to fitting a straight line to the log-transformed counts. The intercept and slope of this regression line are given in the summary() function output above. The curve (with 95% CI) produced by mapping the linear regression line onto the counts can be added to the plot as follows (Fig. 16.2):

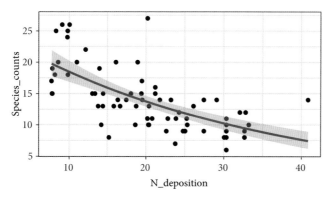

Figure 16.2 Relationship between the number of grassland plant species and nitrogen deposition modelled with a GLM with a Poisson distribution and a log link function.

```
Fig16_2 <-
  qplot(data = Species, x = N_deposition, y = Species_counts) +
  stat_smooth(method = "glm", method.args = list(family = "poisson"))
Fig16_2
```

The curvilinear GLM relationship (linear on the log-y scale) does not predict negative species counts and does a better job of modelling the higher variability at higher species counts. As we've seen, here the dispersion parameter is approximately 1, suggesting the level of variation is as expected if the data were Poisson distributed. If the residual deviance were larger than the residual degrees of freedom (anything above a ratio of about 1.2) but we continued to use the Poisson distribution, then our model would be under estimating the true level of variation in the data (and our standard errors and CIs, etc.). One approach to overdispersion is to use quasi-maximum likelihood.

16.3 Quasi-maximum likelihood

Classical maximum likelihood assumes the level of variability is approximately as predicted for the distribution being used. Switching to quasi-maximum likelihood instead estimates the observed level of variation in the data (as we did earlier with linear models) and adjusts the standard errors accordingly:

```
qml1 <- glm(Species_counts ~ N_deposition, family = quasipoisson,
    data = Species)
```

The estimates of the intercept and slope remain the same, but the standard errors change (of course, here the changes are small because the level of variation is about what we would expect for a Poisson distribution, but if there were marked overdispersion the changes to the standard errors would be larger):

```
summary(qml1)
```

```
##
## Call:
## glm(formula = Species_counts ~ N_deposition, family = quasipoisson,
##       data = Species)
##
## Deviance Residuals:
##     Min       1Q   Median       3Q      Max
## -2.1788  -0.6813  -0.1904   0.5825   3.1749
##
## Coefficients:
##                Estimate Std. Error t value Pr(>|t|)
## (Intercept)    3.210340   0.078168   41.07  < 2e-16
## N_deposition  -0.029436   0.003853   -7.64 7.18e-11
##
## (Dispersion parameter for quasipoisson family taken to be 0.9392865)
##
##     Null deviance: 120.778  on 73  degrees of freedom
## Residual deviance:  64.707  on 72  degrees of freedom
## AIC: NA
##
## Number of Fisher Scoring iterations: 4
```

For interest, we can return to the linear regression model and take a closer look at the residual variation (Fig. 16.3):

```
library(ggfortify)
Fig16_3 <-
  autoplot(lm(Species_counts ~ N_deposition, data = Species),
    which = c(1,3))
Fig16_3
```

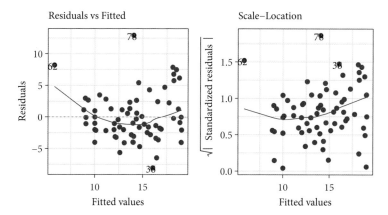

Figure 16.3 Diagnostic plots of the residuals from a linear regression of the count data.

The left-to-right fan shapes in the residuals indicate an increase in the variance as the mean species count goes up, rather than the constant level of variation assumed by the linear regression. However, previously we have tamed these fan-shaped residuals by log-transforming the response and, indeed, log-transforming the species counts does result in acceptable residual plots (try it!). So why bother with the Poisson GLM in this case? As well as taking the integer nature of the data into account, the use of the log link function in the GLM means it is not possible for the model to produce negative values as the linear model would for higher nitrogen deposition levels. The GLM does a better job by modelling the patterns in the mean and variance separately rather than using the one-size-fits-all approach of transforming the data.

16.4 Summary

A GLM with a Poisson distribution is a good place to start when dealing with integer count data. The default log link function prevents the prediction of negative counts, while the use of the Poisson distribution models the variance as approximately equal to the mean. Overdispersion can be diagnosed and dealt with using quasi-maximum likelihood. However, just because the data are integer counts it does not mean the Poisson distribution will necessarily be a good model. The Poisson distribution generally applies well to data with many zeros and low counts. In this example, the level of variation was what would be expected for a Poisson distribution. GLMs using a negative binomial distribution provide one alternative to Poisson GLMs (but are beyond the space limits of this book).

Binomial GLMs

17.1 Binomial counts and proportion data

B inomial counts arise when we have a known number of occasions (binomial trials) on which something of interest could occur and we know how many times that something happened ('successes') and how many times it did not ('failures'). The successes and failures sum to give the *binomial denominator*. Binomial count data can be expressed as proportions (the proportion of successes). However, if the binomial denominator—the size of the trials—varies then it is better to work with the binomial count data because our analysis can weight the larger trials more than the smaller ones rather than treating all proportions equally. However, sometimes the size of the trials is not recorded and we have to work with the proportions.

17.1.1 R PACKAGES

```
library(AICcmodavg)
library(arm)
library(ggplot2)
```

The New Statistics with R: An Introduction for Biologists. Second Edition. Andy Hector,
Oxford University Press. © Andy Hector 2021. DOI: 10.1093/oso/9780198798170.003.0017

17.2 The beetle data

We will begin with a small data set from an experiment looking at the mortality of batches of the flour beetle *Tribolium confusa* exposed to different doses of a pesticide. The data come from a 1935 paper and have been widely analysed since and are included in the AICcmodavg package. We need to use the data() function to load the dataframe:

```
data(beetle)
beetle
```

```
##      Dose Number_tested Number_killed Mortality_rate
## 1 49.06            49             6      0.1224490
## 2 52.99            60            13      0.2166667
## 3 56.91            62            18      0.2903226
## 4 60.84            56            28      0.5000000
## 5 64.76            63            52      0.8253968
## 6 68.69            59            53      0.8983051
## 7 72.61            62            61      0.9838710
## 8 76.54            60            60      1.0000000
```

As well as the numbers of beetles tested and killed, we are also going to want the numbers left alive, so let's calculate those and add them to the dataframe (and, while we are at it, shorten the names of the variables to save space so that we are working with the names tested, killed, and alive):

```
names(beetle)[2] <- "tested"
names(beetle)[3] <- "killed"
beetle$alive <- beetle$tested - beetle$killed
beetle
```

```
##      Dose tested killed Mortality_rate alive
## 1 49.06     49      6      0.1224490    43
## 2 52.99     60     13      0.2166667    47
## 3 56.91     62     18      0.2903226    44
```

```
## 4 60.84       56       28       0.5000000    28
## 5 64.76       63       52       0.8253968    11
## 6 68.69       59       53       0.8983051     6
## 7 72.61       62       61       0.9838710     1
## 8 76.54       60       60       1.0000000     0
```

Eight groups of beetles were exposed to carbon disulphide for 5 hours. The binomial GLM analyses the number of beetles killed (successes for the manufacturer, if not for the beetles!) out of the number tested (the binomial denominator) as a function of the dose of insecticide (the concentration of carbon disulphide in milligrams per litre). When binomial count data are expressed as proportions, the mean must tend asymptotically towards zero (the minimum possible value) and one (the maximum), and this floor and ceiling also constrain the variance. This means we would expect to need some sort of 'S-shaped' relationship to model the mean, while the variance will decrease towards both extremes (0, 1) and be greatest in between. A GLM using the binomial distribution with the logistic curve (the default, canonical link function) models the mean using a symmetric S-shape, while the variance is largest at intermediate proportions and declines towards both zero and one (Box 17.1).

Box 17.1 - Logits and the logistic curve

The logistic transformation converts proportions to logits. Logits are the natural logs of the odds, and the odds are the ratio of successes to failures. If we had a binomial denominator of 10 with five successes to five failures, then the logit would be $\log(5/5) = 0$. You probably won't find yourself thinking in logits very often, but one logit is worth remembering: a proportion of 0.5 is a logit of zero (and, obviously, negative logits correspond to proportions less than 0.5 and positive logits to proportions greater than 0.5). The logistic transformation maps proportions from zero to one onto a symmetric S-shaped curve that asymptotes towards plus and minus infinity (Fig. 17.1(a)). So why not simply use the logit transformation and analyse the resulting logits with a normal least squares regression? Because the variance is not constant: it is larger at intermediate proportions and decreases towards both extremes (Fig. 17.1(b)). Instead, we use a binomial variance function to model the variability.

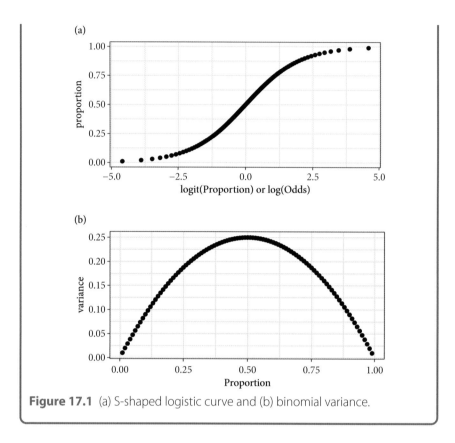

Figure 17.1 (a) S-shaped logistic curve and (b) binomial variance.

17.3 GLM for binomial counts

The GLM of the binomial counts will analyse the number of each batch of beetles killed while taking into account the size of each group (the binomial denominator, the sum of the number killed, and the number alive). Some statistical software packages require the number of successes and the binomial denominator while others—R included—need the numbers of successes and failures. The numbers of dead and alive beetles must be bound together as the successes and failures using the cbind() function so that they can be jointly supplied to the response variable argument for the binomial GLM. We use the family argument to specify the binomial distribution (for which the logistic function—'logit'—is the default link function). The logistic link function and binomial distribution are chosen to take account of the properties of and constraints on the pattern of the mean and variance for binomial count data (Box 17.1). Since we are interested in the mortality rate, we put the number killed as the successes and the number alive as the failures (with apologies to the many beetles out there):

```
m1_logit <- glm(cbind(killed, alive) ~ Dose, data = beetle,
    family = binomial(link = "logit"))
```

The logistic curve is linear on the logit scale, and the coefficients are the regression intercept (−14.6) and slope (0.25) of this line:

```
coef(m1_logit)
```

```
## (Intercept)        Dose
## -14.5780604   0.2455399
```

The CI for the slope supports an increasing probability of mortality as dose increases, as we would expect:

```
confint(m1_logit)
```

```
## Waiting for profiling to be done...
##                     2.5 %       97.5 %
## (Intercept) -17.2645230 -12.1608424
## Dose          0.2056099   0.2900912
```

The display() function output gives the same result in a slightly different form:

```
display(m1_logit)
```

```
## glm(formula = cbind(killed, alive) ~ Dose,
##     family = binomial(link = "logit"), data = beetle)
##             coef.est coef.se
## (Intercept) -14.58    1.30
```

```
## Dose              0.25      0.02
## ---
##   n = 8, k = 2
##   residual deviance = 8.4, null deviance = 267.7
##   (difference = 259.2)
```

This analysis is equivalent to a weighted GLM regression model that analyses the mortality rate as a function of dose using the weights argument to take the size of each group of beetles into account, as we can see if we compare the table-of-coefficients outputs:

```
m2 <- glm(Mortality_rate ~ Dose, data = beetle, family = binomial,
    weight = tested)
summary(m2)
```

```
##
## Call:
## glm(formula = Mortality_rate ~ Dose, family = binomial,
##     data = beetle, weights = tested)
##
## Deviance Residuals:
##     Min       1Q    Median       3Q       Max
## -1.3456   -0.4515   0.7929   1.0422   1.3262
##
## Coefficients:
##                 Estimate Std. Error z value Pr(>|z|)
## (Intercept)  -14.57806    1.29846  -11.23   <2e-16
## Dose           0.24554    0.02149   11.42   <2e-16
##
## (Dispersion parameter for binomial family taken to be 1)
##
##     Null deviance: 267.6624  on 7  degrees of freedom
## Residual deviance:   8.4379  on 6  degrees of freedom
## AIC: 38.613
##
## Number of Fisher Scoring iterations: 4
```

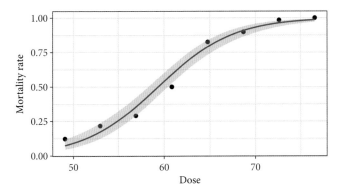

Figure 17.2 Logistic regression with a GLM using a logit link function and a binomial distribution.

The coefficients for the two models are the same. Having shown above that the weighted GLM of the proportions is equivalent to the GLM of the binomial counts, we can use the analysis of the proportions for a graph of the relationship (rather than plotting the numbers alive and dead, it is clearer to plot the mortality rate as a proportion: Fig. 17.2):

```
Fig17_2 <-
   qplot(data = beetle, x = Dose, y = Mortality_rate) +
   labs(y = "Mortality rate") +
   geom_smooth(method = "glm",  method.args = list(binomial), aes(weight = tested))
Fig17_2
```

The summary() function output also allows us to check how well our model meets its assumptions. In particular, as with Poisson GLMs, one assumption of the analysis is that the ratio of the residual deviance to the residual DF (or 'dispersion parameter' in the summary() function output) is approximately 1:1. In this case it is a little higher, as we can see if we look back at the last line of the summary() function output.

This overdispersion is not too substantial, but we can account for it by using the closely related approach of quasi-maximum likelihood. If we alter the family to quasi-binomial then, instead of being assumed (and underestimated), the level of variation will be estimated from the data and the standard errors increased accordingly:

```
m1_quasi <- glm(cbind(killed, alive) ~ Dose, data = beetle,
    family = quasibinomial)
summary(m1_quasi)
```

```
##
## Call:
## glm(formula = cbind(killed, alive) ~ Dose,
##     family = quasibinomial, data = beetle)
##
## Deviance Residuals:
##     Min      1Q   Median       3Q      Max
## -1.3456  -0.4515   0.7929   1.0422   1.3262
##
## Coefficients:
##              Estimate Std. Error t value Pr(>|t|)
## (Intercept) -14.57806    1.46611  -9.943 5.98e-05
## Dose          0.24554    0.02427  10.118 5.42e-05
##
## (Dispersion parameter for quasibinomial family taken
##      to be 1.274895)
##      Null deviance: 267.6624  on 7  degrees of freedom
## Residual deviance:   8.4379  on 6  degrees of freedom
## AIC: NA
##
## Number of Fisher Scoring iterations: 4
```

Notice how the standard errors are increased compared with those from the binomial GLM to take account of the overdispersion. However, one disadvantage of quasi-maximum likelihood is that, because we have a quasi-likelihood rather than a true likelihood, we are not given a value for the AIC (although some statisticians are willing to calculate quasi-AIC values and there are R packages that provide them, including the AICcmodavg package used here).

17.4 Alternative link functions

In some cases, including this one, there is more than one link function that can be used with a given distribution in a GLM. In the case of the binomial GLM, two common alternatives to the logistic function (the default *canonical* link function for this distribution) are the complementary log-log (the wonderfully abbreviated 'cloglog'!) and the probit. We can create alternative versions of model 1 by swapping the logistic link for the probit,

```
m1_probit <- glm(cbind(killed, alive) ~ Dose, data = beetle,
    binomial(link = "probit"))
```

and for the complementary log-log,

```
m1_cloglog <- glm(cbind(killed, alive) ~ Dose, data = beetle,
    binomial(link = "cloglog"))
```

There are various ways in which we could compare these models, but we can take the opportunity to revisit the use of information criteria—here the AIC (taking the sample size as 471, the number of beetles tested)—as explained in Box 17.2.

> ### Box 17.2 - Model selection using information criteria
>
> On the surface, model selection using information criteria is simple—just pick the one with the lowest value. Of course, things are a bit more complicated than that. For one thing, selecting just a single best model has its drawbacks. First, just because a model is the best of a bunch doesn't mean it is any good! Second, there is often a subset of models that are all worth considering. Presenting just a single best model gives the misleading impression that the 'right' model has been found, and hides the uncertainty involved in model selection when finding our way through the complex model selection process (the 'garden of forking paths'). So, it is often better to present a set of two or more models rather than a single best one. But where should we draw the dividing line? Model selection using information criteria tries to reduce some of the issues of arbitrary boundary lines that the levels of statistical significance suffer from and deliberately has fuzzy transition zones rather than hard cut-offs. Of course, this makes it a bit more complex to explain (rigid cookbook recipes are easier to follow). At the one extreme, models within two information criterion units of each other are

essentially equivalent in terms of their parsimony (with an important caveat we'll come to in a moment). At the other extreme, models that are somewhere in the region of 10 to 20 information criterion units worse (i.e. 10 to 20 units higher) than the best model are not usually included in the subset of models to be presented and compared. So, depending on the circumstances, we might consider all models within around 10 to 20 units of the best model. If we are lucky, we may find a convenient gap in the range of the information criterion values (somewhere in the region of approximately 10 to 20 information criterion units) that conveniently separates a leading pack of models from the rest of the field. Now the caveat on models with indistinguishable AIC values (within two units). If two models have similar information criterion values but one is simpler than the other then we might choose to favour the simpler model (all else being equal). In particular, as we saw in Chapter 13, if we have two *nested* models that differ in one term and have information criterion values within two units of each other then we would favour the simpler model. This is because if there is no improvement in the information criterion for the more complex model then there is no support for the additional term (the payback in goodness of fit is not worth the cost of the increased complexity and is similar to what we would get if we added an explanatory variable of randomly generated values). Finally, this book mostly focuses on data generated by experiments that produce nested models. However, much of the literature on the AIC uses examples of non-nested models where different models may contain alternative sets of explanatory variables. The AIC can be used for the analysis of designed experiments, but this may not be where it is of most use. Burnham & Anderson (2002) is the standard reference for more information on the AIC.

Applying the AIC() function to the three candidate models gives their AIC values and their complexity (the number of parameters) as reflected in the degrees of freedom (which we can assign to a dataframe so that we can add some additional information):

```
Cand.models <- AIC(m1_logit, m1_probit, m1_cloglog)
Cand.models
```

```
##                df        AIC
## m1_logit       2  38.61272
## m1_probit      2  37.54547
## m1_cloglog     2  33.83604
```

First, we can subtract the (uninformative) AIC values for the model with the lowest AIC to set that to zero and give the differences in AIC values for the other models:

```
Cand.models$delta_AIC <- Cand.models$AIC - min(Cand.models$AIC)
Cand.models
```

```
##                 df       AIC delta_AIC
## m1_logit     2 38.61272  4.776678
## m1_probit    2 37.54547  3.709433
## m1_cloglog   2 33.83604  0.000000
```

The probit model produces a very small improvement in fit, but models within ~2 AIC units of each other are essentially indistinguishable (all else being equal). The complementary log-log model is ~4 units better, around a fuzzy boundary where we might begin to prefer one model over another. However, all are well within 5 AIC units of each other, easily close enough to be considered as viable alternatives. Rather than select one model, Burnham & Anderson (2002) prefer to perform multimodel inference, basing predictions on a combination of all three models (Box 17.3).

Box 17.3 - Multimodel inference

Picking one 'best' model doesn't reflect the model selection uncertainty involved in arriving at that best model—the journey through the 'garden of forking paths'. An alternative is to work with a set of alternative models. Estimates of the coefficients, and the predictions based on them, can then be an average of this set of models, weighted to reflect the differences in AIC values (see Burnham & Anderson (2002) for details). Of course it is important to think carefully when assembling the set of candidate models to make sure they are all credible alternatives. Multimodel inference also offers an interesting alternative to corrections for multiple comparisons: rather than adjust the P-values for the number of comparisons, the estimates can be a weighted average of the set of models involved in the multiple comparisons (again, Burnham & Anderson (2002) give the details).

In this example, we could see the use of the AIC as overkill because the number of parameters is the same in each case, so that the AIC is only changing due to the change in likelihood—the measures of goodness of fit. I haven't included the log-likelihood values for the models above but it is usually a good idea to do so, so that you can also see this measure of goodness of fit alone, as well as how it combines with the numbers of parameters to produce the AIC values. The output from packages like AICcmodavg will often contain the log-likelihood values for the models being compared, and these values can also be extracted with the base-R logLik() function. But remember, as we saw in the Janka timber hardness regression, fit is not everything—a model that has the best fit may have other drawbacks.

17.5 Summary: Statistics

Binomial count data are often expressed as proportions but it is usually more informative to keep them in count form, where the numbers of successes and failures sum to give the *binomial denominator*, which can be used to give the larger binomial trials more weight in the analysis. With GLMs for binomial count data, the mean is modelled on the logit scale using the symmetric S-shaped logistic curve (although other choices of link function are available), while the binomial variance is largest at intermediate proportions and declines towards both zero and one. Overdispersion can be taken into account using quasi-maximum likelihood. In this case, overdispersion was minor and the analysis allowed us to model the S-shaped response of mortality to increasing pesticide dose using the logistic curve. Alternatives approaches include GLMs using the beta distribution (which are beyond the space limits of this book).

17.6 Summary: R

- In the glm() function for binomial count data, the successes and failures must be bound together using cbind() and jointly supplied as the response variable argument.

17.7 Reference

Burnham, K.P. & Anderson, D.R. (2002) *Model Selection and Multimodel Inference*. Springer.

GLMs for Binary Data

18.1 Binary data

One of the most important uses of GLMs is for the analysis of binary data. Binary data are an extreme form of binomial count data in which the binomial denominator is equal to one, so that every trial produces a value of either one or zero. Binary data can therefore be analysed in a similar way to binomial counts using a GLM with a binomial distribution and the same choice of link functions to prevent predictions going below zero or above values of one. However, despite the use of the same distribution and link functions, because of the constrained nature of binary data there are some differences from the analysis of binomial counts. For one thing, the use of the ratio of the residual deviance to residual DF to diagnose over- or underdispersion does not apply. Given that R's default set of residual-checking plots are also of little if any use when applied to a binomial GLM, this leaves us without any means for model checking with the base distribution of R. Luckily, the arm package, written by Andrew Gelman, Jennifer Hill, and colleagues, provides a graphical approach through the binnedplot() function.

The New Statistics with R: An Introduction for Biologists. Second Edition. Andy Hector, Oxford University Press. © Andy Hector 2021. DOI: 10.1093/oso/9780198798170.003.0018

18.1.1 R PACKAGES

```
library(arm)
library(ggfortify)
library(ggplot2)
library(Sleuth3)
```

18.2 The wells data set for the binary GLM example

Our example data set for a binary GLM comes from an environmental science analysis from Gelman & Hill (2006). The data are not in the arm package but (at the time of writing) can be downloaded from http://www.stat.columbia.edu/ gelman/arm/ and are available in the support materials for this book as Data_Binary_Wells:

```
wells <- read.table("Data_Binary_Wells.txt", header = TRUE)
```

The example concerns an area of Bangladesh where many drinking-water wells are contaminated by naturally occurring arsenic:

```
str(wells)
```

```
## 'data.frame':    3020 obs. of  5 variables:
## $ switch : int  1 1 0 1 1 1 1 1 1 1 ...
## $ arsenic: num  2.36 0.71 2.07 1.15 1.1 3.9 2.97 3.24 3.28 2.52 ...
## $ dist   : num  16.8 47.3 21 21.5 40.9 ...
## $ assoc  : int  0 0 0 0 1 1 1 0 1 1 ...
## $ educ   : int  0 0 10 12 14 9 4 10 0 0 ...
```

   Taking the variables in order, the binary response refers to whether or not people switch the well from which they get their drinking water in relation to the level of arsenic and the distance to the nearest safe well (plus two other social science predictors that we will ignore here).

   One difficulty with plotting binary data is that they obviously all pile up in two rows at the top and bottom of the graph (in any given instance a household either switches wells or it does not; it cannot be somewhere in the middle). Gelman & Hill (2006) add some random noise to jitter the values and spread them out, but a quick alternative is to use a smoother to give us a sense of how the weight of zeros and ones influences the average probability of switching with distance (although one drawback of this shortcut is that the smoother is not constrained to stay within the range of possible values). First, we can define some nicer axis labels:

```
xlabel <- "Distance to nearest well"
ylabel <- "Probability of switching"
```

The default smoother is chosen depending on sample size, so we get a message telling us which method (a generalized additive model, 'gam') was chosen (Wood 2017) (Fig 18.1):

```
Fig18_1 <-
    qplot(data = wells, x = dist, y = switch) +
    labs(x = xlabel, y = ylabel) +
    geom_smooth()
Fig18_1
```

```
## `geom_smooth()` using method = 'gam' and formula
## 'y ~ s(x, bs = "cs")'
```

It looks like the probability of switching declines with distance, as we would expect. Before we fit the GLM, we can avoid inconveniently small coefficient values by rescaling the distance in hundreds of metres:

```
wells$dist100 <- wells$dist/100
```

We can then fit a GLM of switching versus distance, specifying a binomial distribution and the default logit link function, and examine the residuals:

```
fit.1 <- glm(switch ~ dist100, binomial(link = "logit"),
    data = wells)
```

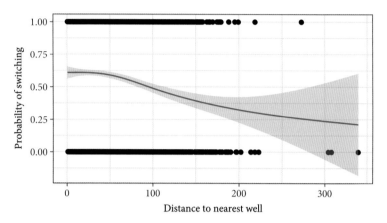

Figure 18.1 Exploratory graph of the probability of switching from a contaminated drinking-water well as a function of the distance to the nearest safe well using the default smoother.

With a normal least squares analysis we could use the automatic diagnostic plots to examine the residuals, but the constrained nature of binary data makes these plots of little if any use (Fig. 18.2):

```
library(ggfortify)
Fig18_2 <- autoplot(fit.1)
Fig18_2
```

As already mentioned, we cannot use the ratio of the residual deviance to residual DF to look for overdispersion as we do with binomial counts (and Poisson GLMs). Luckily, the arm package provides the binnedplot() function, which offers a graphical approach. The data are divided into bins (categories) based on their fitted values and the average residual for each bin is plotted versus the average fitted value. We have to extract the predicted values and residuals and feed them into the binnedplot() function (Fig. 18.3):

```
library(arm)
x <- predict(fit.1)
y <- resid(fit.1)
Fig18_3 <- binnedplot(x, y)
```

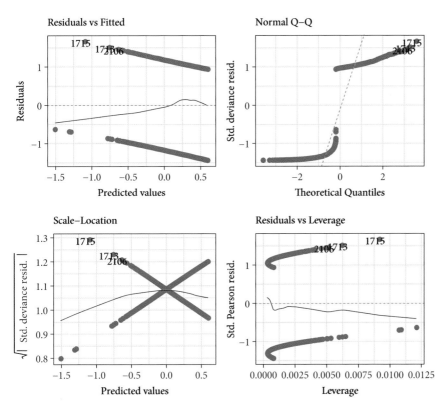

Figure 18.2 The default set of residual-checking plots in R are of little if any use with binary data.

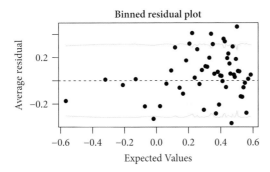

Figure 18.3 The plot of binned residuals produced by the arm package binnedplot() function when applied to the well data.

The grey lines in the plot indicate ±2 standard errors, within which approximately 95% of the binned residuals are expected to fall. Ideally we would like plenty of bins and plenty of values per bin, so the approach works less well for small data sets. In this case, the expectation that around 95% of the residuals fall within the bounds seems to be met, and we can proceed to look at the coefficients,

```
coef(fit.1)
```

```
## (Intercept)        dist100
##   0.6059594   -0.6218819
```

and the confidence intervals,

```
confint(fit.1)
```

```
## Waiting for profiling to be done...

##                      2.5 %       97.5 %
## (Intercept)      0.4882230    0.7246814
## dist100         -0.8140762   -0.4319795
```

It does indeed seem to be the case that the further away an alternative well is, the less likely people are to switch to it. Gelman & Hill (2006) promote a rough rule of thumb for interpreting the slope of the logistic regression: the 'divide by four rule'. Dividing the coefficient for the logistic regression slope by four will give us an approximate estimate of the maximum predicted effect of a unit change in the predictor on the response (given some assumptions detailed by Gelman & Hill (2006)). In this case, a difference in distance of 100 m corresponds to a decrease in the probability of switching of 15% (since $-0.062/4 = -0.15$). The figures show the average probability of switching with a 95% CI (Fig 18.4):

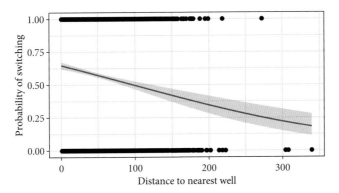

Figure 18.4 Probability of switching as a function of distance to the nearest safe well (in metres) with a 95% CI.

```
Fig18_4 <-
  qplot(data = wells, x = dist, y = switch) +
  labs(x = xlabel, y = ylabel) +
  stat_smooth(method = "glm", method.args = "binomial")
fig18_4
```

We can look at the effect of arsenic concentration in a similar way:

```
fit.2 <- glm(switch ~ arsenic, binomial(link = "logit"),
    data = wells)
display(fit.2)
```

```
## glm(formula = switch ~ arsenic,
##      family = binomial(link = "logit"), data = wells)
##              coef.est coef.se
## (Intercept) -0.31      0.07
## arsenic       0.38      0.04
## ---
##    n = 3020, k = 2
##    residual deviance = 4008.7, null deviance = 4118.1
##    (difference = 109.4)
```

Once again, the estimates are in logits, with a clear positive effect of increasing arsenic concentration on the probability of switching wells, as expected (Fig. 18.5):

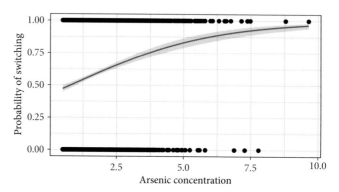

Figure 18.5 Probability of switching wells as a function of arsenic concentration (in micrograms per litre) with a 95% CI.

```
Fig18_5 <-
    qplot(data = wells, x = arsenic, y = switch) +
    labs(x = "Arsenic concentration", y = ylabel) +
    stat_smooth(method = "glm", method.args = "binomial")
fig18_5
```

18.3 Centering

We can also look at the effects of arsenic concentration and distance to the nearest safe well in the same model. Before fitting the GLM, we can make life easier by centering the explanatory variables by subtracting their mean value. This has advantages when a regression intercept of zero is unhelpful or makes no sense (as with a distance of zero metres here—if we took this literally, the new and old wells would be in the same place) and when examining interactions:

```
wells$c.dist100 <- wells$dist100 - mean(wells$dist100)
wells$c.arsenic <- wells$arsenic - mean(wells$arsenic)
```

Considering arsenic and distance at the same time introduces the possibility of an interaction: for a given distance, a household may be more likely to switch the higher the level of arsenic (the model is named fit.5 to match the numbering from Gelman & Hill (2006)):

```
fit.5 <- glm(switch ~ c.dist100 + c.arsenic + c.dist100:c.arsenic,
        family = binomial, data = wells)
display(fit.5)
```

```
## glm(formula = switch ~ c.dist100 + c.arsenic + c.dist100:c.arsenic,
##      family = binomial, data = wells)
##                          coef.est coef.se
## (Intercept)                0.35     0.04
## c.dist100                 -0.87     0.10
## c.arsenic                  0.47     0.04
## c.dist100:c.arsenic       -0.18     0.10
## ---
##    n = 3020, k = 4
##    residual deviance = 3927.6, null deviance = 4118.1
##    (difference = 190.5)
```

The second estimate is for the effect of a change of 100 m when arsenic is at the average level. Similarly, the third estimate is for the effect of a unit change in arsenic for a well at the average distance. As always, an interaction can be viewed from the different perspectives of the variables involved. For every increase of 100 m, we add −0.18 to the coefficient for arsenic. This means that the effect of the level of arsenic declines with distance to the nearest well. Similarly, for every unit increase in arsenic we add −0.18 to the estimate for distance. In other words, the higher the arsenic level, the less important the distance to the nearest safe well.

Once again, we are putting our emphasis on the estimates and intervals rather than the P-values – but what about the significance of the interaction? The size of the estimated difference (−0.18) for the interaction is a bit less than twice the standard error of the difference (0.1), so, while it is marginally significant, it fails to meet the conventional 5% level (as you can explore using the confint() function CIs, the summary() function z-values, or the likelihood ratio test produced by applying the anova() function to a pair of models with and without the interaction). Should we simplify the model by removing the interaction? As we saw in Chapter 13, there is no definitive answer to this question. If it doesn't change the outcome then it doesn't really matter much. In this case, Gelman & Hill (2006) recommend retaining the interaction, given its size and that it goes in the expected direction. Gelman & Hill (2006) explore this in greater detail in their book.

18.4 Summary

We can analyse binary data using a binomial GLM. However, we do not use the ratio of the residual deviance to residual DF to diagnose over- or underdispersion. Instead, we can use the binnedplot() function from the arm package to examine the binned residuals. Otherwise, since the GLM uses the binomial distribution and the logit link function, the analysis is interpreted in a similar way to a GLM of binomial count data. In this case, the GLM of the binary wells data reveals a positive effect of arsenic concentration on the probability of switching wells and a negative effect of distance to the nearest safe well. There is a marginal negative interaction that suggests that each effect slightly reduces the effect of the other (the effect of distance to the nearest well is a bit reduced as arsenic concentration increases, and the effect of arsenic concentration is slightly moderated as distance to the nearest safe well increases).

18.5 References

Gelman, A. & Hill, J. (2006) *Data Analysis Using Regression and Multi-level/Hierarchical Models*. Cambridge University Press.
Wood, S. (2017) *Generalized Additive Models: An Introduction with R*. CRC Press.

Conclusions

19.1 Introduction

This book began with a statistical analysis of the Space Shuttle *Challenger* disaster. Let's close by revisiting that example in the light of the intervening chapters and use it to consider some of the key points of this book.

19.1.1 R PACKAGES

```
library(arm)
library(faraway)
library(ggplot2)
library(Sleuth3)
```

19.2 A binomial GLM analysis of the *Challenger* binary data

The pattern of fuel leaks on previous shuttle launches can be expressed in different ways, including—as introduced in the last chapter—in the form of binary data. The dataframe is called ex2011 and is found in the R package Sleuth3:

The New Statistics with R: An Introduction for Biologists. Second Edition. Andy Hector, Oxford University Press. © Andy Hector 2021. DOI: 10.1093/oso/9780198798170.003.0019

```
str(ex2011)
```

```
## 'data.frame':    24 obs. of  2 variables:
## $ Temperature: int  53 56 57 63 66 67 67 67 68 69 ...
## $ Failure    : Factor w/ 2 levels "No","Yes": 2 2 2 1 1 1 1 1 1 1 ...
```

The dataframe consists of a column of temperatures at the time of previous launches and a factor indicating whether or not there was failure of the O-rings, causing a fuel leak:

```
head(ex2011)
```

```
##   Temperature Failure
## 1          53     Yes
## 2          56     Yes
## 3          57     Yes
## 4          63      No
## 5          66      No
## 6          67      No
```

Data sets sometimes contain errors, as seems to be the case here: in comparison with the graphs in Tufte (2005), an entry for a launch at 63 degrees appears to have been miscategorized as Failure = No. Errors like this should be left uncorrected in the original data set (Wickham & Grolemund 2017), and, instead, following reproducible-research guidelines, any changes made during the analysis should be made and clearly explained as part of the script. In this case, we can correct the fourth entry in the second column to a 'Yes' as follows:

```
ex2011[4, 2] <- "Yes"
```

We can then plot a graph of Failure (indicating whether or not a fuel leak had occurred) versus launch temperature (Fig. 19.1):

```
Fig19_1 <- qplot(data = ex2011, x = Temperature, y = Failure)
Fig19_1
```

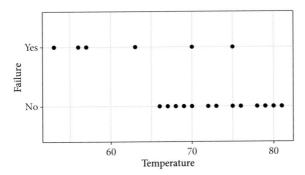

Figure 19.1 Occurrence of failure (fuel leakage) as a function of launch temperature.

Do you think the occurrence of fuel leaks is related to temperature?

Following the approach introduced in the last chapter, we can construct a binomial GLM to perform a logistic regression analysis of the binary data on the occurrence of O-ring failures on previous shuttle launches in relation to temperature. First, we can create a new version of the response (failure, all lowercase) to express the Yes/No coding as binary data (by converting it to a numeric variable of 1s and 2s and subtracting 1):

```
ex2011$failure <- as.numeric(ex2011$Failure) - 1
head(ex2011)
```

```
##    Temperature Failure failure
## 1           53     Yes       1
## 2           56     Yes       1
## 3           57     Yes       1
## 4           63     Yes       1
## 5           66      No       0
## 6           67      No       0
```

The binomial GLM to perform the logistic regression looks like this:

```
m1 <- glm(failure ~ Temperature, family = binomial(link = "logit"),
    data = ex2011)
```

If we extract the logistic regression coefficients and confidence intervals, we can see clear support for a negative relationship between temperature and failure—the chances of a fuel leak were higher in cold weather and decreased as temperatures rose:

```
coef(m1)
```

```
## (Intercept) Temperature
##   14.1700665  -0.2155279
```

```
confint(m1)
```

```
## Waiting for profiling to be done...

##                   2.5 %       97.5 %
## (Intercept)   3.3162485 31.83327461
## Temperature  -0.4724172 -0.05831438
```

We can use a graph to map the logistic regression slope back onto the original scale of measurement as an S-shaped curve that shows how the chance of failure increases as temperatures fall (Fig. 19.2):

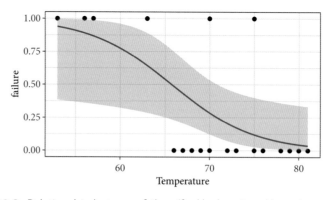

Figure 19.2 Relationship between failure (fuel leakage) and launch temperature.

```
Fig19_2 <- qplot(data = ex2011, x = Temperature, y = failure) +
  geom_smooth(method = "glm",  method.args =
  list(family = "binomial"))
Fig19_2
```

The wide uncertainty, especially at lower temperatures, is reflected in the broad interval but it includes very high chances of failure. The limitation of this graph is that it (cautiously) shows us only the relationship across the range of temperatures for which we have data, but the temperature predicted at launch was around 30°F, far below previous experience. We need to extrapolate from the known relationship to what it predicts at this much lower temperature. Extrapolation is dangerous, as the relationship observed for the restricted temperature range of previous launches may not extend outside this range. Nevertheless, with this caveat in mind, we can use the binomial GLM to make the extrapolation. First, we have only 21 data points, so if we want to take this sample size into account we can use the value of t for a two-tailed 95% CI with $n = 21$, which is slightly above 2:

```
t21 <- qt(0.975, 21)
t21
```

```
## [1]  2.079614
```

Now we can use the predict() function to make predictions from the GLM:

```
predicts <- predict(m1, data.frame(Temperature = 30:85),
    se = TRUE)
```

Next, we calculate a dataframe ('predictions') of values for the curve and for interval upper and lower bounds, using the ilogit function from the faraway package to back-transform from logits to the probability of failure):

```
library(faraway)
fit <- ilogit(predicts$fit)
upper <- ilogit(predicts$fit + t21 * predicts$se.fit)
lower <- ilogit(predicts$fit - t21 * predicts$se.fit)
predictions <- data.frame(fit, upper, lower, Temperature = 30:85)
head(predictions)
```

```
##          fit      upper      lower Temperature
## 1 0.9995493 0.9999999 0.4150330          30
## 2 0.9994409 0.9999998 0.4122349          31
## 3 0.9993066 0.9999997 0.4094173          32
## 4 0.9991399 0.9999995 0.4065783          33
## 5 0.9989333 0.9999992 0.4037160          34
## 6 0.9986771 0.9999988 0.4008283          35
```

Now we can redraw the figure with the extrapolation to the predicted launch temperature (Fig. 19.3):

```
Fig19_3 <-
  qplot(data = ex2011, x = Temperature, y = failure) +
  xlim(30, 80) +
  geom_line(data = predictions, aes(Temperature, fit), colour = "blue") +
  geom_line(data = predictions, aes(Temperature, upper), colour = "darkgrey") +
  geom_line(data = predictions, aes(Temperature, lower), colour = "darkgrey")
Fig19_3
```

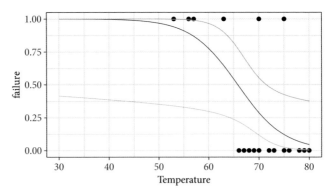

Figure 19.3 Extrapolating the probability of failure to the expected launch temperature (30°F).

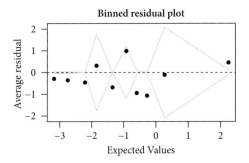

Figure 19.4 An arm package binnedplot() graph of the binned residuals.

According to the prediction from the binomial GLM, the picture just gets worse: our best guess of the outcome is almost certain failure, with even the lower bound of the 95% interval suggesting an unacceptably high probability of failure.

But, we haven't checked how well our model assumptions are met yet. As in the last chapter, we can do that using the arm package binnedplot() function (Fig. 19.4):

```
x <- predict(m1)
y <- resid(m1)
Fig19_4 <- binnedplot(x, y)
```

Recall that the grey lines show a 95% confidence region for the binned residuals. Unfortunately, the majority of the values lie outside this zone. Sadly, the binomial GLM is not a good model for this data, despite it being binary. Other approaches can be tried (Dalal et al. 1989), but none provides an ideal solution for this data set—sometimes that is just how it is. Nevertheless, would the poor residuals lead you to completely disregard the predictions of the model and recommend a launch? Statistics is a fantastically powerful tool, but it is not the only instrument in our toolbox. For example, what do the pre-existing scientific principles and say? Are the predictions in line with them, or are we making extraordinary claims (which may require extraordinary evidence)? As Richard Feynman illustrated during the inquiry using a piece of O-ring and a cup of ice water, the physics predicts the material should become more brittle, supporting the statistical predictions. Ultimately, seven lives were at stake here, and so it would have made sense to use a precautionary approach drawing on all the evidence. Note how we didn't look just at our point estimate—the best guess—but at the whole confidence interval, including the 'best- and worse-case scenarios' of the upper and lower bounds. Here, even an imperfect analysis might have saved the seven astronauts when properly considered together with the underpinning science.

19.3 Recommendations

In this short final section, I reiterate what I see as some of the most important points from this book in the form of 10 parting recommendations (reflecting my own current preferences):

1. *Make your research reproducible.* The R Markdown package now makes it very easy to produce reproducible research documents directly from R.

2. *A picture is worth a thousand words.* Graphs are often the best way to present your results to the reader (you can always present the numbers in supplementary tables and R Markdown documents). As well as presenting the results using figures, use graphs to explore and understand the data during the analysis and model building, and to assess how well the assumptions of the analysis have been met. In the case of the *Challenger* disaster, a good graph alone might have prevented the ill-judged decision to launch.

3. *Keep it simple.* In most cases, try not to leap immediately to a complex analysis. Take time to explore and understand the data and to estimate simple statistical summaries (estimates and intervals) that give you a sense of the effect sizes and their uncertainties. Where appropriate, use preliminary simple analyses as a basis for model building. This has multiple advantages. In particular, if your all-singing, all-dancing final model fails to converge after

a month of hard work, you are not left with nothing to show for your efforts.

4. *Consider more than one model.* Stepwise multiple regression is coming under increasing criticism. Often there is more than one model that deserves consideration. This model selection uncertainty goes unseen when only a single 'best' model is presented. New approaches using information criteria provide one framework that leads more naturally to the consideration of multiple models.

5. *Attempt the P-free challenge.* One of the main messages of this book is to follow the current statistical advice to focus less on *P*-values and statistical significance and more on the estimates and effect sizes. The scientific significance of the results is much more important than the statistical significance (statistically significant results may be scientifically unimportant, especially when sample sizes are large and power is high). In our research group, we have sometimes attempted to write papers using only estimates and intervals. I don't think we have published a *P*-value-free paper yet; usually some make an appearance, if only in response to reviews. But, despite the statistical shortcomings of these papers (I am sure there are many), the *P*-values that do appear should be there for a reason and not just to blindly follow convention. You might also want to consider the recent recommendation by Wasserstein and colleagues (and many others) to abandon statistical significance entirely (see Chapter 11).

6. *Report estimates, intervals, and sample sizes.* One simple thing that would probably improve statistical reporting in the life and environmental sciences more than any other would be the regular use of confidence intervals. If papers reported as many estimates and intervals (together with sample sizes) as *P*-values, things would be dramatically improved. In addition to benefiting the primary reporting of results, it would also greatly facilitate meta-analysis.

7. *Go back to basics.* In order to use and report intervals, we have to understand them, but I am constantly amazed by how much work employing complex approaches does not seem to understand these basics (no doubt I still have a lot more to learn myself). Make sure you understand the various types of error bars and intervals and how to interpret them, including in graphical form.

8. *Make a focused plan of analysis.* Some of the factors contributing to the reproducibility crisis could be avoided by making a focused plan for an analysis before beginning.

9. *Give P-values the respect they deserve (and no more).* In some ways we take P-values too seriously, and in other ways not seriously enough. By this I mean that we sometimes take P-values at face value, a value they may not deserve in the light of our sloppy analysis. The P-value depends on many things. Ideally, we would know in advance exactly how we were going to do the analysis (including any transformations and so on) so that a test could be done once and once only, with all assumptions met and so on. Sadly, these conditions are rarely met, so that, to some degree or other, we end up with mushy Ps. On the other hand, if we are going to use P-values, we should try and make them as meaningful as possible by satisfying as many of the conditions that they depend on as we can.

10. *Focus on repeatability, not small P-values.* No matter how small the probability value (or how high the level of confidence), a result can always be a false positive. With so many scientists doing so many analyses, it happens all the time. This is part of the process; it doesn't mean we have done anything wrong. It is only human to leap to the defence of our hard work when it is challenged, but scientific results only become established after they are shown to be repeatable. One of the lessons of the reproducibility crisis is that science prizes novelty too highly, to the detriment of establishing a solid foundation of truly repeatable results.

19.4 Where next?

The first (2015) edition of this book ended with some further extensions
of linear models, with a chapter on mixed-effects models (MEMs) and
another on generalized linear mixed-effects models (GLMMs). Sadly, the
GLMM example stopped converging with later versions of the software.
That, together with page limits and the opinion of some reviewers that
mixed-effects models could not be introduced in a single chapter, meant
that those two chapters have been dropped from the second edition of
this book to make way for some new additions. However, you can find my
materials for these two chapters in the online support material (see below).
MEMs are also covered in some of the books cited in earlier chapters,
particularly Gelman & Hill (2006) and Maindonald & Braun (2010).

19.5 Further reading

- The most up-to-date popular book introducing statistics is *The Art of
  Statistics* by David Spiegelhalter (2019).
- *Significance* magazine is a great place to get an overview of the
  history, development, and current state of statistics. This journal is
  a revamped version of the old *Bulletin of the Royal Statistical Society*
  and they have done a superb job—it is very accessible and readable.
  It's the only journal I subscribe to!
- Highland Statistics is a group that consults, teaches, and publishes
  books on statistics using R, often aimed at ecology and related
  subjects.

19.6 The R café

My working title for this book (and one of the courses it was based on)
was *Contemporary Analysis for Ecology*, abbreviated to 'café'. Sadly, the
publishers had other ideas, with a view to wider marketing. However,

I instead opened the R café on the PlantEcol.org website, which I share with Lindsay Turnbull (Associate Professor in the Department of Plant Sciences at the University of Oxford). The R scripts, data files, corrections, and supporting materials will be available there (as time and resources allow)—see you at the R café!

19.7 References

Dalal, S.R, Fowlkes, E.B., & Hoadley, B. (1989) Risk analysis of the space-shuttle: Pre-Challenger prediction failure. *Journal of the American Statistical Association* 84: 945–957.

Gelman, A. & Hill, J. (2006) *Data Analysis Using Regression and Multilevel/Hierarchical Models*. Cambridge University Press.

Maindonald, J. & Braun, J.W. (2010) *Data Analysis and Graphics*. Cambridge University Press.

Spiegelhalter, D. (2019) *The Art of Statistics*. Penguin.

Tufte, E. (2005) *Visual Explanations*. Graphics Press.

Wickham, H. & Grolemund, G. (2017) *R for Data Science*. O'Reilly.

A Very Short Introduction to R

OUP already has two books that focus on introducing the R software, and *Getting Started with R* by Beckerman et al. (2017) and Insights from Data by Petchey et al. (2021) are a great place to begin if you are new to the language. The focus of this book is linear-model analysis, so this final chapter gives only a very brief introduction to R tailored to its use in this book. To be frank, I debated whether it was worth including this very brief introduction to R, given the huge amount of freely available material on the internet (see below). However, I suppose if this book is your reading for a long-haul flight (or is it only me who takes stats books?), as you take the enforced offline time to finally learn how to do analysis in R, then this final chapter should provide at least a minimal introduction to get you started until you are back online. There are countless videos online offering support for R and RStudio. Perhaps the single best book-length resource is *R for Data Science* (Wickham & Grolemund 2017), which is available online for free as a website (https:// r4ds.had.co.nz/).

20.1 Installing R

One of the great advantages of using R is that it is free to download and install, giving you greater freedom to work on your own machine

The New Statistics with R: An Introduction for Biologists. Second Edition. Andy Hector, Oxford University Press. © Andy Hector 2021. DOI: 10.1093/oso/9780198798170.003.0020

or any other computer where you have the rights to install software. R is installed from CRAN, the Comprehensive R Archive Network (http://cran. r-project.org). The installation is usually simple, but if you have problems then an online search almost always quickly provides a solution. Once R is installed and you have activated it by clicking the R icon you are confronted with the R console. A short block of introductory text gives a few details about R, including the version of R that you are running. Underneath this text is a flashing cursor following the '>' prompt. You can work interactively in R by typing commands at the cursor and entering them. Depending on what you type, R will display a response on screen or, if no output is required, will simply return a new cursor ready for your next instruction. However, if you work in this way you will lose your work when you quit R. Instead, it is better to open a second window and to type your commands there instead. These can be sent to the console with a simple keystroke (Command + R with Mac OSX, Control + R with Windows). The advantage of this approach is that you can save the text typed into this window as an R script (a text file with the extension .r or .R) that records your entire workflow (all of the instructions you give to R), which you can save and reuse. Rather than write every R script from scratch, you can often open an old one as a starting point and edit it for the new task (remembering to give it a new name!). In this way you will quickly build up a collection of R scripts for different purposes and can exchange them with other R users, who can run your analyses for themselves (in a way that collaborators using two different commercial software packages cannot). Working with R scripts also means that you have a record of your work and can easily repeat every step of the process. One of the advantages of this is that it allows you to document data checking, correction, processing, and manipulation. I advise you to use spreadsheets as little as possible—just to enter the data and record metadata that are essential for understanding the data themselves (units etc.). Then, use R for all subsequent steps (as I was revising this chapter, it emerged that COVID-19 cases had been underestimated due to an error linked to the use of Excel). For example, you can use R to check for errors in the data, to correct these errors and save a new version of the data, to process the data (e.g. converting from one unit to another), and for data manipulation (rearranging the data, taking subsets of data, and so on). If you work in this way, you will have a program that records every stage of your work from data entry, through the analysis, and all the way to production of the final figures for publication. However, rather than working directly in R, this book uses a second piece of software—RStudio—which provides an interface that is identical on all platforms (Mac, PC, and Linux) and which has a lot of supporting features built in.

20.2 Installing RStudio

RStudio can be installed from https://rstudio.com/products/rstudio/download/#download and is usually trouble free. When you open RStudio, you will see that is has a console window that displays the same introductory text as in the R console. However, RStudio has four panes: the console, a source panel (where R scripts and R Markdown documents open), and two additional ones to organize graphs, files, and R packages (see below). The screenshot in Fig. 20.1 shows RStudio on my laptop as

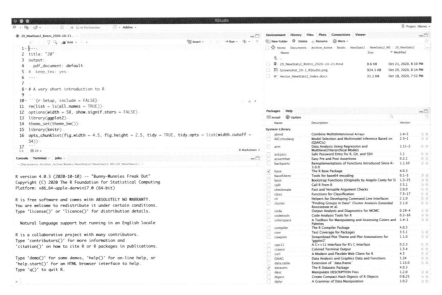

Figure 20.1 RStudio displaying all four panes.

I write this chapter. The top left panel displays the R Markdown file for this chapter (see Chapter 4 for more on R Markdown), with the console at the lower left (displaying the opening R blurb) and the panes on the right-hand side showing a list of files and a list of R packages. The panes on the right have multiple tabs showing other information. When in RStudio, it is best to write R code into a new R script or R Markdown document rather than typing into the console directly (and losing it when you quit). The RStudio Code menu gives several choices for running single lines or regions of R code (together with their keyboard shortcuts). When a line of code in the script or R Markdown document is run, it appears in the console window.

20.3 R packages

When you download R, you install a 'base' version which performs core functions only. However, thousands of add-on packages can be downloaded from CRAN (under 'Software' on the left-hand side) that supply additional functionality. It might be helpful to think of CRAN as the app store for R packages. The packages used in each chapter of the book are given at the start of each chapter (which is the recommended way to write R scripts and R Markdown files). Packages contain functions, a key feature of the R language (see below). Packages can also contain useful data sets and other resources. You can install (and update) packages using the RStudio Tools menu. Once installed, packages have to be activated, which can be done by ticking the box in the packages tab (bottom right in Fig. 20.1) or using the library() function.

20.4 The R language

20.4.1 FUNCTIONS

Functions are a key element of the R language. R functions have names and end in a pair of empty parentheses. For example, as we have noted, the library() function activates packages for use. Most of the chapters in this book start by activating all of the packages they require (I nevertheless often redundantly include the library() function again at first use to indicate where the package is needed). For example, if we want the darwin data, we need to activate the SMPracticals package that contains them (and to install that package if we have not already done so):

```
# install.packages('SMPracticals', dependencies =
# TRUE))
library("SMPracticals")
```

To avoid reinventing the wheel, one package can use functions in another—it is dependent on them (for these actions). Missing packages are a common cause of error messages. To avoid this, when installing packages always download any others that they depend on too (by ticking the relevant menu box or, as shown above, by typing 'dependencies = TRUE' inside the parentheses). If you still get an error message referring to missing packages then installing that package explicitly usually fixes things.

20.4.2 ARGUMENTS

A second feature of the R language is the arguments that are given inside the function parentheses, separated by commas. In the install.packages() function call above, the first argument is the name of the package to be installed (SMPracticals—note that it needs to go in double quotation marks and, if using a text editor, be careful it has not substituted these for two singles or a format of double quotes R does not recognize!) and (as we have seen) the second instructs R to install any additional packages that are required. Missing commas between arguments are another common cause of errors. Many arguments have binary TRUE/FALSE alternatives.

20.4.3 OBJECTS

A third feature of the R language is that it is an object-oriented programming language. In this case, the darwin data set is an object of class data.frame (R lingo for a data set):

```
class(darwin)
```

```
## [1] "data.frame"
```

The str() function tells you that the darwin data set contains four columns (variables) and 30 rows:

```
str(darwin)
```

```
## 'data.frame':    30 obs. of  4 variables:
## $ pot   : Factor w/ 4 levels "I","II","III",..: 1 1 1 1 1 1 2 2 2 2 ...
## $ pair  : Factor w/ 15 levels "1","2","3","4",..: 1 1 2 2 3 3 4 4 5 5 ...
## $ type  : Factor w/ 2 levels "Cross","Self": 1 2 1 2 1 2 1 2 1 2 ...
## $ height: num   23.5 17.4 12 20.4 21 ...
```

Variables (columns) within data sets can be referred to using the dollar sign, as follows (only the first six values are shown):

```
darwin$height
```

```
## [1] 23.500 17.375 12.000 20.375 21.000 20.000
```

Objects can be created in R using the assignment arrow (a dash followed by the 'greater than' or 'less than' symbol, depending on which way you want the arrow to point). For example, the mean() function can be used to get the average of the 30 values in the column named height in the darwin data set and save the answer as an object called 'average_height' (obviously you could give it any name, probably a shorter one, but be careful to avoid using a name already used for something else in R!):

```
average_height <- mean(darwin$height)
average_height
```

```
## [1] 18.88333
```

Arguments inside functions have to be either in the expected place (positional matching) or explicitly named (in which case position does not matter). Naming is a bit more typing (but not much, given RStudio suggestions and copy–paste) but is clearer and then we don't have to worry about matching the expected order:

```
qplot(data = darwin, y = height, x = type)
# Plot not shown to save space
```

R code can be annotated using the hash symbol, #. On a given line, text following a hash symbol is ignored by R. You can also use a leading hash to 'turn off' following bits of R code while retaining them in case you want to reactivate them later by deleting the hash.

20.4.4 DATAFRAMES

Most of the data sets used in this book are available as part of an R package (as a so-called dataframe) but in some cases the data are loaded from a file (a .txt or .csv file). Specifying paths is error prone and complex (your directory structure is different from mine). The simplest way to work is to put your data file in the same folder as the R Markdown file that you are using to run the analysis, as this is where R will look for the data, and you therefore need to specify only the file name (as shown in Chapter 12).

One potentially confusing thing is how to tell R where to find the data you want to work with. Perhaps the simplest option is to use functions that have an argument specifically designed for this. For example, both the lm() function and the qplot() function have the 'data =' argument (as shown in the last chunk of R code). However, not all functions have this argument. There are then a variety of ways to proceed. The first is to specify the dataframe and variable name using the dollar sign, like this: dataframe$variable. A second option is to 'attach' a dataframe using the function of the same name, use the data for the task in hand, and then detach it (using detach()). However, this can be error prone, especially for beginners. A third, quicker option is to use the with() function, which attaches and detaches with less typing (with this dataframe do this . . .). You will see all three approaches widely used, so all three are also used in this book at different points—you can pick how you prefer to work.

20.4.5 GRAPHICS

R has at least three widely used graphics systems—the base-R functions (plot() etc.) and the lattice and ggplot2 packages. The ggplot2 package is the most advanced and is being developed further into the interactive ggvis software. I have therefore chosen to focus on that. However, the ggplot2 syntax is the least familiar (lattice is more similar to base-R code). I have therefore used the quick-plot function, qplot(), as much as possible, as it allows the most sophisticated plots with the shortest, simplest code—it used to be possible to generate panel plots with, say, regression lines and confidence intervals in one relatively simple line of code. It now takes a bit more code to achieve this in the current version of ggplot2, but I think it is still the best compromise.

One of the challenges of learning analysis is that it means getting to grips with both the statistics and the software being used to implement it. I have tried to strike a balance that keeps the R code relatively short and simple, especially at the start of the book. I hope this very short introduction will be enough to allow offline readers to understand the basics of R so that they can focus on the applied details of the statistical analysis that are the real focus of this book.

Index

-> assignment ('gets') arrow 19, 256
* asterix character 48, 153
: colon character 153
$ dollar character 23, 256
# hash character 30, 257
() parentheses character 13, 254
> R prompt 252
[] square (indexing) brackets 115, 148
~ tilde character 52, 203

A

Akaike information criterion (AIC) 158, 170
AICc 170
AICcmodavg package 218, 224, 227
analysis notebooks 32
analysis of deviance 204
analysis of variance (ANOVA)
    one-way 127, 141, 143, 149
    factorial 139, 141, 148, 152
    sequential 156–157, 174, 184, 186
    table 108, 130–134
anova() function 130, 157, 204
arm package 52, 60, 111, 232
attach() function 257
axis labels 83, 165

B

Bayesian statistics 5, 133
binnedplot() function 232, 245
binomial
    count data. See data
    distribution See distribution
    denominator 217–220
blocks See experimental design
box-and-whisker plot 20

boxcox() function 20
Box-Cox family of transformations 200

C

cbind() function 220
centering 78, 236
Chi-squared 170, 201
coefficient 53, 55
coef() function 86
coefplot() function 60, 116, 144, 150
comparison
    multiple 145–147, 227
    pairwise 127, 138, 145–147
Comprehensive R Archive Network, CRAN
        37, 252
confidence interval. See interval
confidence levels 73
confint() function 58–61, 78, 151
confounding effects 123
contrasts
    a priori 6, 102, 147
    treatment 130
correlation 75
count data See data
covariate 162, 183–184
CRAN. See Comprehensive R Archive Network

D

data
    argument in R functions 19, 22
    binary 229
    binomial count 217
    count 209
    entry 253
    indexing 115, 148

data (*cont.*)
  manipulation 253
  sets 2, 28
data() function 73
data frames (analysed)
  beetle 218
  darwin's maize 16
  janka 73
  case1402 162
  grassland biomass (ANOVA) 140
  grassland diversity (poisson GLM) 210
  wells (drinking water) 230
degrees of freedom (DF) 131
detach() function 257
deviance 170, 204–5
dispersion parameter 213, 223
display() function 52, 56
distribution
  binomial 219–220
  Chi-squared 170, 201–204
  F 134
  Gaussian, *See* normal distribution
  gamma 205
  Poisson 209–212
  normal 45–46
  t 98–10
divide-by-four rule of thumb (logistic
    regression) 234

E

effect size 44, 148, 158
error bars 26, 108–116
estimate 44
estimation-based approach 5, 97, 107,
    118–121, 151
expand.grid() function 110–111
experimental design 5, 28, 137
  balanced 156, 164, 178–180
  blocked 179
  factorial 140–141, 164
  paired 66
explanatory variable. *See* variable

F

F-value 132–135, 190
F-test 108
faceting 166
factor 17
factorial *See* experimental design
false positive 44, 127, 146

fitted values 63, 79
fitted() function 79

G

general linear models 162, 172, 186
generalized linear models 4, 162, 195
geometric objects ('geoms') 20, 26
ggplot2 package 26, 90, 93

H

help() function 36
heteroscedasticity 82
homoscedasticity 82
head() function 10, 16

I

inference 43
  multi-model 227
information criteria 170, 225
installation (of R) 252
install.packages() 255
interaction 152–158
interaction.plot() function 152
intercept 52–56
interval
  confidence 42, 60, 90
  prediction 119
  profile 201, 207

J

Jittering 54

K

k (number of parameters). *See* AIC

L

lattice package 258
labs() function 160
least significant difference (LSD) 117–120
least squares 84
  iterative weighted 201
library() 11, 254
likelihood
  surface 201–202
  ratio 237
  ratio test 237
linear predictor 203, 209
link function 203
  identity 204–205

log 209, 213
  logistic 220, 225
  square root 205
lm() function 4, 52
logistic
  curve 201–228
  regression 234
logits 219

M

main effect 153, 156, 158
margin of error 49
marginality 186
MASS package 68, 200
matrix 92
maximum likelihood 201
mean() function 22
mean squares 22, 132
median 22
meta-analysis 6, 122
metadata 252
model(s)
  additive 5, 231
  building 246
  complexity 170
  factorial *See* experimental design
  selection 225, 227
  simplification 171–172
model (X) matrix 87
model.matrix() function 88

N

normal distribution. *See* distributions
normality 64
null hypothesis (expectation) 44, 60
null hypothesis significance testing 107, 121, 124

O

odds 219
orthogonality 179, 189
options() function 134
outliers 41, 66
overdispersion 223–224

P

panel plots 258
parameter(s) 45, 53, 78, 170
parsimony 170, 226
penalty 170
percentile 59

pf() function 134
point estimate(s) 44, 49
power (statistical) 134, 146, 153, 192, 247
predict() function 87, 91–92
prediction 85–94, 119
predictor 73
probability (P) values
  criticism of 5, 133, 247
  definition 133

Q

qplot() function 11, 90, 166, 258
qt() function 101, 243
quasibinomial family 224
quasipoisson family 214
quasi-maximum likelihood 213, 177

R

R café (webpage) 6, 249
R
  language 254
  Markdown package 32
  scripts 30, 250
r-squared 53, 78
randomization 137
read.table() function 140
regression
  equation 86
  intercept 72
  polynomial 198
  slope 73–78, 234
  stepwise multiple 247
relevel() function 62, 147
repeatability 123, 248
replication 179
reproducibility 12, 29
residual variation 63, 72, 79, 84, 91
residuals 79–82
residuals() function 80
response 18

S

sample size (n) 43, 134, 192, 231
scripting 29
sd() function 23
signal-to-noise ratio 108, 128, 132
significance
  statistical 32, 133, 146, 247
  biological (practical) 98
  levels 49, 118

Significance (magazine) 249
simplicity 172
sleuth package 19, 162, 164
smoother, statistical 76, 198, 231
SMPracticals package 16
sqrt() function 24
Standard error
    of a mean (SEM) 42–46, 56, 62
    of a difference (SED) 57–58
standard deviation (SD) 23–24, 43–47
statistics
    Bayesian 5, 133
    frequentist 5, 48, 133
    information theory 170
    new 6
str() function 17
subsetting 190
subset() function 143
sum of squares (SS) 128, 130, 157, 170, 182, 202
summary() function 17, 102

T

t-test 97
t-value 99
transformation 196, 200
    and interactions 165

log 165, 201
square root 196

V

var() function 23
variable
    confounding 123
    explanatory 28, 82, 226
    response 166, 179
variance (*see also* mean squares) 23
    components 133, 137, 190
    function (in GLMs) 203
    pooled estimate of 63–64, 132
    ratio 133

W

weights argument (glm) 222
with() function 22–23, 75, 257
workflow 36, 222

X

xlab() function 84

Y

ylab() function 84